Interactive Technologies and Music Making

Challenging current music making approaches which have traditionally relied on the repetition of fixed forms when played, this book provides a new framework for musicians, composers, and producers wanting to explore working with music that can be represented by data and transformed by interactive technologies.

Beginning with an exploration into how current interactive technologies, including VR and AR, are affecting music, the book goes on to create an accessible compositional model which articulates the emerging field of 'transmutable music.' It then shows how to compose and produce transmutable music for platforms like video games, apps and interactive works, employing tutorials which use a range of inputs from sensors, data, and compositional approaches. The book also offers technical exercises on how to transform data into usable forms (including machine learning techniques) for mapping musical parameters, and discussion points to support learning.

This book is a valuable resource for industry professionals wanting to gain an insight into cutting edge new practice, as well as for assisting musicians, composers, and producers with professional development. It is also suitable for students and researchers in the fields of music/audio composition and music/audio production, computer game design, and interactive media.

Tracy Redhead is a musician, composer, and researcher working in music performance, production, and technology. Her career spans from singer/ songwriter roots to pioneering interactive music projects with cutting-edge technologies. Holding a Master of Arts and PhD in Music, she is currently a lecturer in Electronic Music and Sound Design at the University of Western Australia. She explores the intersection of science and art, and novel forms of expression.

Interactive Technologies and Music Making

Transmutable Music

Tracy Redhead

Routledge
Taylor & Francis Group

LONDON AND NEW YORK

Designed cover image: @ Getty

First published 2025
by Routledge
4 Park Square, Milton Park, Abingdon, Oxon OX14 4RN

and by Routledge
605 Third Avenue, New York, NY 10158

Routledge is an imprint of the Taylor & Francis Group, an informa business

British Library Cataloguing-in-Publication Data
A catalogue record for this book is available from the British Library

ISBN: 978-1-032-22651-4 (hbk)
ISBN: 978-1-032-22650-7 (pbk)
ISBN: 978-1-003-27355-4 (ebk)

DOI: 10.4324/9781003273554

Typeset in Times New Roman
by MPS Limited, Dehradun

Access the Support Material: www.routledge.com/9781032226507

Contents

Figures

Tables

Acknowledgements

I have so many people to thank during my 15-year obsession with Transmutable Music.

Special thanks to Prof Richard Vella for reading and providing feedback on this book. Also, thank you for your guidance as my principal PhD supervisor.

Hannah Rowe for encouraging me to start this book and Lucy McClune for getting the proposal over the line.

Thanks to all my colleagues and students at the University of Western Australia Conservatorium of Music for your guidance and support. Special thanks to Dr Chris Tonkin, Ass Prof Sarah Collins, and Ass Prof Ionat Zurr.

Thanks to Dr Florian Thalmann, Prof Julian Knowles, Ass Prof Jon Drummond, Hakan Lidbo, Ginger Leigh, Trevor Brown, Dr Kristefan Minski.

The community of gorgeous people in Linz, Austria, my home away from home.

Sofie Loizou, for the inspiration to begin this journey and the support during it!

The FAST project, Prof Mark Sandler, Dr Adrian Hazzard, The Global Music Tech Fest community, Prof Andrew Dubber and Michela Magas, The Ars Electronica, and all the artists inspiring, changing, and challenging the world.

Thanks to all my friends for helping to balance life by adding fun and happiness, and also for your support, directly and indirectly.

Big thanks to my family, the Redheads and the Rutherfords, for all your continued support. Especially my Mum and Dad, who have always given me all the love, support, and confidence I could have ever needed.

Finally, my heart goes out to my husband, Jonathan Rutherford, for all his support, love, and patience. A true companion on adventures afar!

1 Introduction

Waves of Change

Change. It is a concept that can both intrigue and intimidate us. We strive to create stability and secure environments for ourselves and our loved ones, often resisting the inevitable waves of transformation surrounding us. However, change is an intrinsic part of our existence – a constant force that shapes our reality and the world we inhabit.

In this book, I will attempt to explore the profound relationship between music and the ever-evolving landscape of interactive technologies. I delve into the depths of how music, as an art form, embraces interactivity, and adapts to the digital and data-driven age. This exploration highlights a paradox that exists within recorded music's realm – the industry's historical embrace of new technologies, yet its hesitance in fully integrating emerging interactive products and platforms inspired by the gaming industry.

I pose a fundamental question: How can we bridge the gap between music and interactive technology? Moreover, we explore how interactive technologies can pave the creation of new musical forms and serve as mediums for expressing ground-breaking artistic concepts and ideas.

Before embarking on this journey, let us ponder the timeless words attributed to Heraclitus (535 – 473 BC), "No man ever steps in the same river twice, for it is not the same river and he is not the same man."[1] Although philosophised centuries ago, these concepts continue to resonate with us, transcending gender and human-centric perspectives.

Let us expand the scope of Heraclitus' interpreted wisdom to encompass all beings that engage with a river, regardless of gender or cellular composition. For every living being that interacts with the flowing water, both the river and the being undergo constant change and transformation.

The river itself becomes a metaphor for change – an entity in perpetual motion. Its currents shape the world beneath its surface, altering the very foundations on which it flows. Fish swim through its depths, while frog larvae evolve into tadpoles. The river represents the ebb and flow of life – a testament to the inherent nature of change.

DOI: 10.4324/9781003273554-1

Now, let us shift our focus to the being within Heraclitus' quote. Consider a human, a complex organism predominantly composed of water. Within this biological structure, microscopic mites, and beneficial bacteria reside upon the skin, blood courses through veins, and consciousness interprets the sensory inputs of the world. Each passing moment brings forth new experiences and transformations, whether immersed in the river or beyond its banks.

Returning to our analogy, envision a section of the river where the water stagnates. The absence of flow leads to an unpleasant odour, a deterioration that parallels aspects of our lives. Stagnation impedes growth and poses challenges when change inevitably arrives. Yet, we strive to create safe environments that resist change, even though we can only delay it for a limited time.

Change, as we know, can be inextricably linked with the passage of time. If change is constant, why do we, as a society, cling to the familiar? The reason I commence this book with a philosophical quandary is that the ensuing chapters challenge preconceived notions about music consumption and experience. These challenges trace their roots back to the very essence of change. This book is a tribute to music's transformative nature, exploring how it can be moulded and reimagined through the manipulation of data.

Music and Technology

Music and technology have historically intertwined to shape musical innovation, from the Neanderthal flute and the printing press to digitisation. This symbiotic relationship means that music is perpetually evolving alongside technological advances. In the digital era, amid revolutionary changes in communication and industry practices, the music industry finds itself in an extended transition, grappling with the sluggish evolution of copyright, royalties, and equitable compensation for artists. All this whilst rushing towards the irresistible and transformative potential of AI and blockchain technologies. Music has historically been at the forefront of transformative technological advances.

This state of flux is not unique to the digital age. Historically, music has adapted to new technologies, laws, and cultural shifts. For example, the introduction of radio into households sparked fears among venues and record labels that live performances and record sales would plummet. Why would people pay for music if they could listen to it for free on the radio? Initially, radio broadcasts did not compensate musicians despite earning advertising revenue. It took nearly 15 years to establish the copyright and royalty systems we know today.

Considering the significant time it took for the broadcasting and music industry to develop a fairer remuneration system for musicians, it is unsurprising that the sweeping changes brought on by digitisation, the internet, and communication advancements present a considerable challenge. It is clear that there is a substantial journey ahead in modernising our copyright and royalty frameworks to accurately reflect contemporary consumer behaviour and artist practices.

A significant challenge of digitisation is the exact replication of digital copies to the original, a stark contrast to the past, where each subsequent copy of music would lose quality. For instance, the 30-second mashups I produced as a teenager using my cassette recorder never reached professional quality due to this degradation. However, a 30-second mashup created today can achieve the highest quality due to digital clarity.

The high quality of digital audio resulting from digitisation has created many opportunities to innovate all things music. Previously, the sounds produced by musical instruments were constrained by their physical design. A guitar, for example, could offer strumming, individual notes, or percussive sounds by tapping on its body. Today's electronic music production tools, including samplers, sequencers, and digital audio workstations (DAWs), transform any recorded sound into a potential instrument. The modern laptop serves as a platform for software that crafts unique digital instruments. These can incorporate audio samples from any source, be it another artist's work or sounds recorded by the user themselves, expanding the creative possibilities beyond traditional limitations.

In today's post-digital era, the music industry grapples with evolving issues around copyright, royalties, and income streams, yet the creation and distribution of music are flourishing. Internationally, industry professionals, artists, and academic teams are innovating at the intersection of music and technology. There's a burgeoning interdisciplinary synergy as music merges with computer science, neuroscience, artificial intelligence, and engineering, all amidst shifts in consumer culture. This fusion is creating a dynamic environment ripe for innovation. Musicians are now integrating music production with coding and user experience design, paving the way for fresh musical experiences and novel ways of emotional expression that are keenly awaiting discovery.

Music distribution, too, has been transformed. Traditionally, music was a tangible commodity, treasured by many as part of a physical collection of albums, singles, or EPs distributed on mediums like CDs, vinyl records, or tapes. This 'ownership model' of music consumption reigned supreme for the better part of the last century. However, as highlighted by Wikström in 2012, we have shifted to an 'access model' – characterised by the emergence of streaming services and online subscriptions such as Spotify and Apple Music. Wikström posits that the value generated from recorded music now pivots more on the context – how music is experienced – rather than merely owning a copy (Wikström, 2012).

This paradigm shift has seen the perceived value of recorded music plummet, not just in financial terms. There's a sentiment, echoed in music industry circles, that recorded tracks have become akin to business cards – tools for exposure rather than saleable items. Despite the ease of recording music today, the artistry, time, and soul invested in creating original and compelling music often go undervalued. Consequently, musicians are exploring innovative ways to infuse value into their work, responding to audiences' craving for more engaging and personalised musical experiences (Baym and Burnett, 2009).

Yet, as we anchor ourselves in this post-digital landscape, we encounter another disruption: the advent of AI technology. Digitisation has not only revolutionised production tools and consumption patterns but has also sparked growth in the music tech sector and an upsurge in the number of music creators. It has also reignited debates around copyright and royalties. However, the true potential of digitisation remains untapped. Pfeiffer and Green (2015) remind us that when music is digitised – transformed into data – it becomes malleable, opening up new avenues for manipulation and creativity (Pfeiffer and Green, 2015).

Despite these changes, music's essence, its ability to reflect, critique, and deconstruct the world, endures. Yet, the traditional, static presentation of music – rooted in analogue technology – is due for a renaissance, spurred on by the innovations in interactive technology. As we contemplate the future of music, it's crucial to revisit and learn from past technological milestones that have shaped its evolution.

Recorded Music and Static Formats

During the recording industry's century of world domination, it was focused on the process of recording music and making copies of it. The analogue technologies underpinning this remarkable innovation profoundly impacted how society perceived and understood music at that time.

The recording industry has evolved significantly, transitioning through various distribution and playback mediums such as tapes, records, CDs, MP3s, mini-discs, downloads, and vinyl. The first format being the phonograph record invented in 1877 and gaining popularity around 1912.

With each new medium, not only did the manufacturing of the medium itself expand, but the accompanying playback devices also flourished. From record players and CD players to hi-fi systems and radios, technological advancements brought new ways for people to experience music.

The rise of the internet and digitisation altered the landscape dramatically. The once-thriving market of CD player manufacturers, record labels, and record stores has dwindled, giving way to the dominance of Internet Service Providers (ISPs), streaming platforms, and computer/phone manufacturers within the music industry. Technology companies like Spotify, Apple, Microsoft, and YouTube currently dominate the music industry. Streaming services and digital platforms have become the primary means of music consumption, leading to significant changes in revenue streams for artists.

Digitisation, coupled with the evolution of the internet, has transformed our interaction with music – how we consume, produce, and distribute it. Yet, despite these advancements, the fundamental structure of a song or track remains static; music products such as albums and singles have continued to be distributed as fixed recordings, even though the tools for playback and production have fully embraced digital technology.

'When you think about it, recorded music is something of a historical oddity. Imagine telling a visitor from 1650 that we like to listen to the exact same performances, over and over again, without any variation ... they might well be baffled. Or, as Brian Eno once said, perhaps it is our grandchildren who will be the baffled ones' (Steadman, 2012).

Brian Eno's reflection on recorded music as an anomaly in history suggests that future generations may find our repetitive listening habits curious (Steadman, 2012). This perspective is rooted in the history of music and existed long before the advent of recording technology.

The dawn of the printing press marked a pivotal moment for the music industry, which came to revolve around music publishing companies profiting from sheet music sales. Engaging with music traditionally meant attending live performances, ranging from full orchestral renditions to intimate gatherings where eager musicians wielded their instruments of choice. Western historical vignettes often depict homes resonating with the collaborative sounds of families and friends, frequently congregating around a piano. At this time, music was something that you had to physically interact with; it was a verb.

Live performances of a song are inherently unique, influenced by factors such as the venue, audience, and the emotions and experiences of the musicians. In contrast, recorded music presents a paradox: it offers a singular, preserved performance that remains unchanged with each listen.

The concept of fixed music playback, now a norm, traces back to the emergence of analogue recording. This breakthrough allowed for the transient artistry of live performances to not only be recorded but also conserved and duplicated for broader public consumption. Such recordings – once mixed, mastered, and replicated – offered a static representation of music. They encapsulated a singular, unalterable performance in time. Often, these recordings captured the synergy of well-practised musicians in harmony, enshrining what many would call a magical moment – a snapshot of a truly extraordinary musical event.

As recording technologies evolved, so too did the techniques for capturing music. The introduction of multi-track recording and overdubbing allowed recorded music to stitch together disparate moments in time, seamlessly blended with the help of studio techniques like reverb and compression. This innovation meant musicians no longer needed to be in the same room – or even the same country – to create a cohesive musical piece. Despite the geographical dispersion of talent, the final recordings remained static in form, whether destined for vinyl or digital streaming platforms like Spotify. Today's recordings largely adhere to the traditional approach of immortalising a moment in music. Despite technological advances, the essence of arrangements and production remains tethered to the musical frameworks established by earlier analogue technologies.

Importantly, static versions of recorded music evoke nostalgia. Bands are often judged based on their live performance's fidelity to the recorded track as if a song has only one 'correct' form of rendition. This entrenched notion of a song's static form has become deeply rooted in our understanding of music, challenging our ability to envision beyond it.

Reflecting on the philosophical quandary introduced at the start of this chapter, the link between human nature's resistance to change and the preservation of music in its static form becomes apparent. We strive to capture time and savour moments, similar to how we hold onto memories of life's experiences. This parallel draws into focus the current state of recorded static music, which seems anchored in our collective nostalgia. I confess that certain recordings of songs resonate deeply with me, anchoring specific times and cherished events in my life, symbolising a desire to maintain constancy in an ever-changing world. Music, however, is intrinsically a temporal art; it is perceived differently across cultures, reflecting diverse conceptualisations of time in their musical expressions.

Our perception of time influences how we experience a static recording of a song. Despite the recording being static, it can elicit different experiences based on the listener's mindset. For example, I might enjoy a track's mix one moment, feeling it sounds perfect. Yet, if I play the same track for someone else, my perception may shift to criticism, influenced by the anticipation of the other person's judgement. This phenomenon illustrates the contrast between subjective and objective listening, a variance in perception I am sure we all have encountered.

Time encompasses both duration and spatial dimensions. Bergson (1869–1941) pondered time's multiplicity in his philosophical discourse on duration. He differentiated between the quantitative aspect – measurable, metric time akin to clock ticks – and the qualitative – time as a subjective experience (Bergson, 2014). Phrases like "time flies when you're having fun" encapsulate this duality.

If time's duality can be understood as comprising both measurable and experiential qualities, then music, too, can be appreciated through this dual lens.

Despite initial resistance due to his mathematical background, Bergson insisted on time's qualitative nature. Deleuze clarified that Bergson did not aim to dispute Einstein's theory but rather to illuminate the subjective variance in temporal perception based on identical events (Deleuze, 1991).

Music listening can alter time perception, allowing listeners to become immersed and lose track of chronological time. Conversely, observing a clock or any temporal metric can slow perceived time by juxtaposing measured time against spatial experience. Music, while adhering to its own rhythmic and metrical constructs, can transcend specific durations in subjective experience.

Cultural backgrounds significantly influence time perception. Kramer theorised that Western culture's time perception is goal-oriented, reflected in music that often builds towards a climax or resolution. Conversely, cultures like Bali's, which view time cyclically, produce music such as Gamelan that eschews linear resolution and embraces cyclical structures (Kramer, 1988).

McTaggart differentiated between static time – a linear timeline – and dynamic time – a flowing stream of events (Kaae, 2008; Faye et al., 2000). In this context, static recorded music aligns with a timeline, a structured form designed to unfold from beginning to end. Dynamic or Transmutable Music, however, exists within a time stream, with its elements not fixed to a linear structure but free to manifest within a fluid temporal continuum.

This static concept of music is now deeply ingrained and does not truly reflect its potential as a dynamic and interactive art form. Technological progress in sampling, automation, DSP, AI, and multi-tracking has revolutionised the recording process. Studios and laptops have become instruments themselves, opening doors to endless creative possibilities. Yet, the music industry still clings to the traditional presentation of recorded music – why?

The initiation of interactive technologies challenges this static paradigm. As emerging tools redefine music's potential, the industry must consider how new forms, influenced by interactive elements in composition, performance, and audience interaction, can reshape musical experiences.

As history informs us, static and fixed music works are not the only way music can be perceived and experienced. Future 'technology native' generations are already demanding new musical experiences. Undoubtedly, static recorded versions of music are still important and relevant, but they emerged and were influenced by the technology of the time and pushed by industry to increase revenue. Given that interactive technologies are the emerging technologies of our time, how could they influence new recorded music forms? What happens if a musician introduces an interactive element into the composition, performance and/or audience role?

This idea that music is fixed in time or the same each time it is performed has become an accepted understanding of experiencing recorded music. Changing this perception challenges the foundation of music and the society it represents. Throughout history, changing a musical form has never happened quickly as the form of popular music represents the societal system we live in with its regulations, copyright and media (Sinnreich, 2010). However, we find ourselves in a paradox. If music is now digitised (resulting in the loss of much of its monetary value) and can be represented by data, it has the potential to be entirely transformed by interactive technologies.

Interactive Technologies and the Shaping of New Music Experiences

Interactive technologies are now part of our everyday lives. Interactive technologies include but are not limited to mobile phones, wearables, Internet of Things (IoT), sensors, video games etc. Simplistically, interactive technologies allow for a two-way communication. For the technical explanation, "Interactive technology allows for a two-way flow of information through an interface between the user and the technology; the user usually communicates a request for data or action to the technology with the technology returning the requested data or result of the action back to the user" (IGI Global, 2020).

A good example of an interactive technology is the telephone. Two people can communicate in real-time. Another way to think about interactive technologies is passive versus active, linear, and non-linear. Watching a movie has a start and an end; therefore, this would be an example of passive or linear media. Of course, you can interact by crying, laughing or discussing the movie, but your actions will not affect how the movie is played back. The only available transmutable interaction is limited to being able to pause, play, stop, record, and adjust the TV display settings or the volume of the movie. An example of active or non-linear media would be a video game, where the player's actions can change the game's playback. The player is interacting with the game and creating their own personalised experience. Obviously, interactivity is not something new that has come about with technology but rather something that technology has been trying to emulate. Humans and other animals have been communicating with each other the whole time, and it is communication that is at the centre of interactive technologies.

Music has always been an interactive experience – a dynamic exchange between the performer and the listener, manifest in actions such as singing, playing an instrument, or even just the act of attentive listening. Focusing on the digitised and recorded forms of music, we explore how these can be crafted to incorporate interactive qualities.

Since the early 1990s, we have seen commercial music artists explore interactive concepts, as seen in Fred Davis's (1993) WIRED article "I Want My Desktop MTV." Davis speculated a seismic shift in both the public's engagement with music and the industry's financial framework due to emerging interactive technologies. He observed an eagerness among major media entities to invest in the nascent CD-ROM technology, which promised new interactive musical experiences (Davis, 1993).

Fast forward thirty years or so, and the evolution of interactive music continues, yet the initial promise of a transformed musical landscape remains only partly realised. The anticipated revolution in interactive music has not fully materialised, prompting the question: why has this promising field not reached its full potential?

The quest for engaging musical experiences persists, with the music industry continually seeking innovative methods to captivate audiences and generate revenue. From the CD-ROMs of the 90s to today, we have seen a remarkable transition. The landscape now includes Web 2.0, Web 3.0, Semantic Web, mobile devices, the Internet of Things (IoT), wearables, Extended Reality (XR), sophisticated music production tools, audio middleware, and advancements in AI. The pace of technological advancement shows no sign of slowing, ensuring a perpetually expanding toolkit for the industry.

Back in 1993, the challenge was compatibility – ensuring that interactive content could be widely accessed. Today, the gaming industry has largely resolved this through adaptive and interactive music models. Moreover, music software has dramatically progressed, offering sophisticated tools for production and interaction, exemplified by the continued development of

DAWs, Max, M4L, RNBO, KONTAKT, Pure Data, and advanced gaming audio middleware.

This exploration delves into interactive technologies as they pertain to human-system interaction. This encompasses gaming and web applications and trends such as VR, AR, album apps, mobile apps, gesture sensors, and wearable tech.

As an art form, music naturally lends itself to interaction. Integrating interactive music could, arguably, offer audiences a deeper, more enveloping experience than traditional recorded formats. Today's songwriting and production processes utilise lyrics, melodies, harmonies, rhythms, and performances to convey a wide array of sounds and emotions. Interactive technologies offer a new palette of tools to augment and reshape the listening experience, fostering new forms of musical engagement and reinventing the ways in which stories and ideas are conveyed.

Interactive technology's potential impact on popular music is further underscored by its prominence at major industry events like SXSW, MIDEM, and The Great Escape. As musicians increasingly leverage new tech to connect with their audience and find novel revenue streams, interactive music products are on the rise. However, there seems to be a disconnect: the vast potential of interactive methods in music creation is recognised, but their adoption is slow, with the music industry and many artists hesitating to fully embrace these emerging platforms, unlike their counterparts in the gaming sector.

To envision how popular music could evolve alongside interactive technologies, one must first grasp the various facets of music creation: composition, performance, production, and distribution. These processes have traditionally shaped music into fixed forms, from blues and rock n' roll's 12-bar and 32-bar structures to the verse-chorus and anatomy prevalent in rock, pop, and electronic dance music. Despite these conventions, variation remains a cornerstone of music's appeal.

Young suggests that video games, a domain where non-linear music thrives, could pave the way for broader adoption of such music forms. The concept of 'active listening,' where listeners actively curate their musical experience, mirrors the dynamic systems used in gaming and could herald a new era for music consumption.

Despite this visionary outlook, the music industry, alongside its audience and artists, shows resistance to change. There are differing opinions on whether music should be a malleable experience or preserved as the artist intended.

This tug-of-war between innovation and tradition presents a formidable challenge to those navigating the intersection of interactive technology and music. Nevertheless, it is clear that music has the potential to become as fluid and multifaceted as the technologies that could shape its future.

Pfeiffer and Green (2015) underlined the fundamental shift in music production brought about by digitisation, stating that audio must be understood as data that can be manipulated (Pfeiffer and Green, 2015). This philosophy underpins the myriad of data sources, from physical gestures and geo-location information to biometric and environmental inputs. All can be intricately woven into the fabric of sound creation. The implications for music production are profound; every interaction, from a swipe on a touch screen to the rhythm of a heartbeat, becomes a potential input for musical expression. Allowing for recorded music to truly represent the intrinsic changing nature of music performances.

Creating a music system that can adeptly map and transform this diverse data is now at the heart of modern songwriting and production. Navigating this new terrain requires artists to engage with complex data sets, often calling for advanced tools like machine learning algorithms and AI, to distil and apply these inputs meaningfully within musical compositions.

Artists are increasingly exploiting these innovations, integrating mobile apps, XR, VR, AR, and other emergent technologies to forge new sonic landscapes. However, developing such interactive music products necessitates a collaborative synergy among musicians, technologists, and researchers – a multiplicity of roles (Holland et al., 2013). The Interactive Audio Special Interest Group (IASIG) further emphasises that delivering interactive audio is a multifaceted endeavour, necessitating a concerted effort from a wide range of specialists.

The surging interest in data-driven music is reshaping the entire landscape, yet there is an evident disconnect between the proliferation of advanced interactive technologies and the prevailing practices of popular music production. This book proposes the term 'Transmutable Music' as an umbrella concept to encompass music influenced by data. This term captures the essence of soundscapes that fluidly respond to various data inputs, in stark contrast to static playback methods.

As we traverse the junction where technology meets creativity, there is a palpable opportunity – and necessity – for the music industry to align with the dynamic capabilities afforded by modern interactive technologies. The task ahead is multifaceted, involving the development of music systems that can interpret and apply a vast array of data, the creative integration of these systems into interactive platforms, and the challenge of composing within this Transmutable context. It is an era where music is not just heard but interacted with, creating an immersive experience that continually evolves in tandem with listener engagement and environmental context. This evolving paradigm calls for a concerted effort to redefine music production in harmony with these transformative technological advances.

Overview of the Book

This book examines the field of Transmutable Music, a term I prefer over the more common 'dynamic music.' I will explain my reasoning for this change in Chapter 2. At its core, it introduces a compositional model designed to adapt and respond to data, reshaping how we perceive musical creation and interaction.

Section 1 Understanding Transmutable Music Systems

We begin by exploring the aesthetics of Transmutable Music, defining key terms like music form, structure, and process, and discussing how they relate to this innovative music form. This foundational understanding is crucial for grasping the concepts that follow.

A pivotal aspect of Transmutable Music is its relationship with time and change. We examine this interplay, highlighting how music both occupies and is perceived in time, and dissect the components that constitute a Transmutable Music system.

The heart of the book is the detailed exploration of these components: experience, content, musical architecture, and control systems. This section provides a framework for understanding music form and structure within Transmutable Music and offers insights into musical design approaches.

Section 2 Approaches for Composing Transmutable Music

Moving into practical applications, I present three methodologies for designing and composing Transmutable Music: experience, variability, and transmutability. Each approach is dissected through case studies, user testing analyses, and practical exercises, focusing on evaluating a Transmutable Music system.

First, this section explores experience, interaction design, and evaluation techniques like user studies for Transmutable Music Systems. The following chapter provides approaches to composing with variability.

Case studies and tutorials have been included to enhance your understanding of these concepts. They cover various methods like variation, layering and blending (vertical orchestration, branching (horizontal resequencing), and algorithmic approaches. These practical approaches aim to solidify your grasp of composing with variability.

The concept of transmutability is further explored, examining how data can alter musical form. I discuss the relationship between music and data, explore music representation models, and provide historical context for autonomous and algorithmic music systems.

In the final chapter of this section, an evaluation model for Transmutable Music is presented. It aims to help composers and producers evaluate their own practice.

Section 3 Interactive Technologies and Music Making (Background and Context)

Section 3 provides some context by exploring the evolving relationship between music, technology, and audience participation. Chapter 8 traces the journey of recorded music from its historical roots as a passive form of consumption to its current state as a vibrant, interactive culture. It dissects the transformative role of remixing and its progression from 'musique concrète' to the digital age, where technology like DAWs and new formats are revolutionising how we interact with and compose music. The chapter also navigates through the complexities brought about by AI, copyright laws, and licensing, hinting at a future where music composition is increasingly influenced by the interactive demands of video gaming.

Chapter 9 builds on this foundation, outlining interactive art and its influence on participatory music experiences. It explores the concepts introduced by notable figures such as Joel Chadabe and Brian Eno. The chapter outlines the rise of 'Mobile Music', Album apps and web-based interactions that have expanded the scope of how we experience music. It describes the pathway where future recorded music is an active, engaging experience, propelled by the listener's interaction and shaped by their environment. Together, these chapters weave a narrative of music's transition into a domain where audience engagement is as integral to the experience as the sound itself.

Section 4 Compositional Approaches in Transmutable Music: Tutorial Series

This practical-based section is made up of eleven tutorials and two creative exercises. Supported by the Companion Website, the tutorials are designed to be a dynamic learning experience. Whether you are new to Transmutable Music or looking to refine your expertise, these resources cater to various skill levels and interests. Transmutable Music encompasses multiple disciplines, including music composition, production, computer science, data science, interaction design, and user experience. The tutorials are crafted to be accessible, avoiding the need for in-depth knowledge in any single area.

Chapters 10 and 11 feature six tutorials on 'Mobile Music Making.' These are designed to comprehensively introduce fundamental mapping techniques, MIDI and OSC protocols, data mapping, modelling, machine learning, and user experience design.

In Chapter 12, you'll find 'Adaptive Music for Video Games' tutorials. This area provides a robust framework for integrating music with data. These tutorials use audio middleware (*FMOD*) to guide you through composition and data mapping methods, including effectively managing transitions and control parameters. Game development platforms like *Unity* are also explored, demonstrating how to create systems for prototyping music playback.

A significant challenge in composing Transmutable Music is the ability to experience the music fully within its intended system, and these tutorials aim to bridge that gap.

Section 5 Conclusion, Companion Website, and Glossary

The book culminates with a look toward the future of Transmutable Music, considering its potential and evolution. As we come to the end of this book, we look ahead and consider the future of Transmutable Music. Chapter 13 will encourage you to imagine what may be possible in this exciting field.

For a list of support materials, including videos, links and tutorials, Chapter 14 provides an overview of the Companion Website, with details on how to access it.

Sometimes, the book can get a little dense with technical terms and in-depth concepts. Chapter 15 provides a detailed glossary with key definitions to help guide you through.

How to Use This Book

Please feel free to navigate the book in any way that works best for you. Each section offers an opportunity to learn something new, no matter where you start.

For those who want to find more resources, Chapter 14 provides a guide to the Companion Website. Here, you will find a variety of supporting materials, such as videos, links, and tutorials to enhance your learning experience. This companion is the key to accessing additional content.

We know that the book deals with complex topics and technical language, so Chapter 14 includes a comprehensive glossary. You will find clear explanations of key terms to help you easily navigate the book's more intricate concepts.

Software

In this book, I will utilise several standard software including *Max 8.5, M4L, Ableton Live,* and *Touch OSC*. Most of these programs offer free trial versions, and I have provided links for introductory courses in the tutorial sections in case you are unfamiliar with any of them. The Companion Website will also list links to software, trials, and special discounts. Please note that there is a nominal fee for *TouchOSC*, which is around $20 for the mobile app (In 2024).

Designing tutorials for a diverse audience with varying levels of familiarity with these software tools is a challenge. However, rest assured that the principles outlined in this book are versatile and can be applied using different software platforms. For instance, while *Pure Data (or PD extended)* offers the capability to execute most tasks we discuss in this book, I have specifically chosen Max due to its seamless integration with *Ableton Live* through *Max for Live (M4L)*.

So, join me as we traverse the exciting landscape where music and interactive technologies intertwine, with the hope of inspiring new artistic expression and innovation.

Summary Points

Music has consistently evolved with technology, from ancient instruments to digitisation.

The digital era challenges copyright and royalty systems. Digitisation allows perfect copies, contrasting the quality loss in analogue duplication.

Digital tools enable new creative potentials beyond traditional instrument limitations.

There has been a transition from ownership to access model in music consumption. Recorded music's value shifts from possession to the context of experience.

Musicians are seeking innovative methods to add value to their work in response to changing consumer behaviour.

Digitisation of music opens doors for creativity through data manipulation.

Despite technological advances, music remains largely static in its presentation.

The future of music must revisit past technological milestones to understand its evolution.

The recording industry's influence on music perception has been profound.

Technological shifts have altered music consumption but not the fundamental static nature of songs.

Interactive technologies allow two-way communication and are ubiquitous in modern life. The music industry explores interactive music but faces challenges in adoption.

There is a potential disconnect between the capabilities of interactive technology and current music industry practices.

Data-driven music, termed 'Transmutable Music,' represents a shift towards dynamic, responsive soundscapes.

The music industry must embrace modern interactive technologies to meet evolving consumer expectations.

Collaboration among musicians, technologists, and researchers is key to developing new Transmutable Music systems.

Note

1 It should be noted that Heraclitus's well-known and accepted quote is an interpretation of his original concept.

References

Baym, N, and R Burnett. 2009. *Amateur Experts: International Fan Labour in Swedish Independent Music*. London, ROYAUME-UNI: Sage.

Bergson, Henri. 2014. *Time and Free Will: An Essay on the Immediate Data of Consciousness*. London, UK: Routledge.

Davis, Fred. 1993. "I Want My Desktop MTV." WIRED. 1993. https://www.wired.com/1993/03/desktop-mtv/.

Deleuze, Gilles. 1991. *Bergsonism*. Translated by Hugh Tomlinson and Barbara Habberjam. New York: Zone Books.

Faye, Jan, Uwe Scheffler, and Max Urchs. 2000. *Things, Facts and Events Edited by Jan Faye, Uwe Scheffler and Max Urchs*. Amsterdam: Atlanta, GA. Rodopi.

Holland, Simon, Katie Wilkie, Paul Mulholland, and Allan Seago. 2013. "Music interaction: understanding music and human-computer interaction." *Music and Human-Computer Interaction*, edited by Simon Holland, Katie Wilkie, Paul Mulholland, and Allan Seago, 1–28. London, Heidelberg, New York, Dordrecht: Springer.

IGI Global. 2020. "What Is Interactive Technology | IGI Global." 2020. https://www.igi-global.com/dictionary/parental-mediation-of-adolescent-technology-use/41845.

Kaae, Jesper. 2008. "Theoretical Approaches to Composing Dynamic Music for Video Games." In *From Pac-Man to Pop Music: Interactive Audio in Games and New Media (Ashgate Popular and Folk Music Series)*, edited by Collins Karen, 75–91. England: Ashgate Publishing Company.

Kramer, Jonathan D. 1988. *The Time of Music New Meanings, New Temporalities, New Listening Strategies*. New York: Schirmer Books.

Pfeiffer, Silvia, and Tom Green. 2015. *Beginning HTML5 Media: Make the Most of the New Video and Audio Standards for the Web*. New York: Apress.

Sinnreich, Aram. 2010. *Mashed Up: Music, Technology, and the Rise of Configurable Culture*. Amherst, MA: University of Massachusetts Press.

Steadman, Ian. 2012. "Brian Eno on Music That Thinks for Itself." http://www.wired.co.uk/article/brian-eno-peter-chilvers-scape: WIRED. 2012.

Wikström, Patrik. 2012. "A Typology of Music Distribution Models." *International Journal of Music Business Research* 1 (1): 7.

Section 1

Understanding Transmutable Music Systems

2   Aesthetics of Transmutable Music

Dynamic or Transmutable Music?

Imagine a world where recorded music is no longer a static entity but a dynamic, changeable form that interacts and changes, just like a live performance does. This is not a far-off fantasy but a reality emerging in 'Transmutable Music.' This term represents a new frontier of music creation and experience.

Introducing an appropriate term or framework to describe fluid music can be challenging. Throughout my years of trying to compose such music, I have grappled with the word 'fluid' – a descriptor that does not quite capture the nature of the process, which is far from seamless.

Subsequently, I gravitated towards 'Dynamic Music,' a term that featured prominently in my PhD thesis. However, the ambiguity of this nomenclature proved potentially confusing to some peers, given the existing connotations of 'dynamics' in music and music production. Nonetheless, I still believe 'dynamic' aptly embodies, be it generally, the essence of music that is responsive and adaptable to data.

The music industry frequently uses buzzwords like 'interactive' and 'dynamic' to market music products or discuss music-centric businesses such as Spotify or YouTube. However, while these labels are used, the music itself is not inherently dynamic or interactive. It may possess interactive elements such as social features, the ability to reorder songs, or the periodic release of new tracks from an album, but the music remains static.

For instance, an artist might launch an interactive app for fans to engage with the artist or their concept, but the music within does not adapt or react to data. In 2016, the International Federation of the Phonographic Industry (IFPI) coined the term 'Dynamic Album' to describe tracks released over a lifespan, such as weekly releases (IFPI, 2016). But the music released still exists in a fixed form, hardly meeting the definition of dynamic.

I propose using 'Transmutable Music' to circumvent such confusion with the ambiguity of the term dynamic. As a result of my PhD studies and in consultation with my PhD supervisor, Prof Richard Vella, the term 'transmutable' surfaced. This term is appealing because of its novelty. However, I cannot

DOI: 10.4324/9781003273554-3

shake off the feeling that it does not perfectly fit the concept I want to express. Be it 'Dynamic' or 'Transmutable' music, I will venture to delineate and discuss my use of the terms. My use of transmutable is much more specific than dynamic.

'Dynamic Music' is often employed in game audio and academia to characterise music that adapts to data. The concept of something non-static over time is inherently dynamic. Other terms like interactive, adaptive, reactive, responsive, and algorithmic, along with contextual, are often utilised by musicians and artists. However, these terms can sometimes need more precise or universally agreed-upon definitions.

The term 'Transmutable' is new within musical terminology. 'Transmute' essentially means to change the form of something or 'to alter from one state to another' (Etymonline, 2023). Merriam-Webster (2023) states, 'Transmutation involves the transformation of one thing into something else.' In the context of music, data is used to transmute or change the music from one form to another.

As I interpret it, Transmutable Music is an all-encompassing term encapsulating various forms of music that can react, adapt, and respond to data or information. This vast category could include a range of interactive, reactive, dynamic, adaptive, generative, algorithmic, AI-driven, autonomous, and contextual music forms. This list, while extensive, is undoubtedly incomplete as creatives continually coin new terminologies to capture the music they aspire to create. Transmutable Music unites music from various fields, including game audio, interactive art, computer music, and generative music. The data can transform the form, structure, organisation, creation, and overall mix of sound or musical material. Consequently, this data becomes integral to the composition or production process.

My motivation for formulating this overarching term was to propose a compositional model to assist composers and musicians in navigating and operating in these evolving fields. However, to comprehend what Transmutable or Dynamic music is, it might be helpful to understand what it is not: it is neither fixed nor static.

Regardless of the genre - classical, electronic, pop, rock, or soul – if a piece of music is composed as a fixed entity, it would not fall into the categories of Dynamic or Transmutable Music. Indeed, these forms could be dynamic in their approach to song structure, timbre, or the build-up to a crescendo. This can lead to some confusion around the term 'dynamic.' A song can possess dynamic elements yet still maintain a static form. Transmutable Music is, however, fluid or dynamic in form, and its creation process is far from straightforward. It cannot be interpreted using traditional concepts of music form.

Exploring these ideas has occupied more than a decade of my life, resulting in a method for composing music that can morph and adapt in real time. The compositional model I will introduce in this section is the culmination of a mind-boggling journey through a labyrinth of ideas and theories. Often, my discussions about Transmutable Music have been met with dismissal or scepticism. Critics argue that fixed songs are central to our musical understanding, fostering love for music and an irreplaceable sense of nostalgia associated with specific recordings.

In response to such reservations, I do not advocate for a complete overhaul or replacement of traditional music with Transmutable (or Dynamic) music. Instead, I view this as a new branch, a pathway to an innovative method of composing and experiencing music. Given the historical interplay between technology and music, it is hard to deny the transformative potential of these new interactive technologies. They may not substitute the traditional music experience, but they will unquestionably contribute to reshaping it.

So, the terms 'Transmutable' or the more general 'Dynamic' music represent my vision for new and emerging music forms interacting with data.

In this section, I will present a model for composers and musicians that provides a framework illustrating how data can play a central role in the compositional and production process. Because Transmutable Music is music that can adapt and change according to data, it can create new experiences for audiences and artists.

While I have proposed this umbrella term, I also feel it is essential to define different forms of Transmutable Music:

- **Interactive** – The user directs the interaction, having real-time control over audio and visuals.
- **Adaptive** – Mostly employed in gaming, it refers to non-linear music composed to adapt to and support gameplay and user actions.
- **Autonomous** – This system-led music operates independently, including generative, AI compositions and systems.
- **Reactive** – This music reacts to data or an environment where the data stream is continually updated, e.g., in a scenario where car driving controls the playback.
- **Responsive/Contextual** – This music adapts to the user's environment or actions, but not through direct control. The user's contextual data, such as location or weather, influences the music's playback.
- **Generative** – Music created by a system constantly evolving.
- **Algorithmic** – This type of music is generated using mathematical approaches, including Markov chains, stochastic algorithms, automata, and intricate Fourier analysis.
- **AI** – A subset of Algorithmic music that uses learning, problem-solving, or knowledge-based algorithms to create music, including machine learning and deep learning algorithms.

To summarise, Transmutable Music offers a listening experience that goes beyond a static form. It can be created through either direct or indirect actions, which can be initiated by humans or machines. Regardless of whether the form and structure of the music is generative, algorithmic, or adaptive, there are similarities in how data influences its form, process, structure, and overall experience. The music reacts to the data in some way, creating a foundation that all music fields can build upon.

To explain the compositional model for Transmutable Music, I first need to address how I use temporality, form, and structure of music within a dynamic construct.

Music Form and Structure

Understanding the correlations between form, structure, and time is crucial when composing Transmutable Music.

The concept of form in music can be interpreted differently, depending on various factors such as musical style, genre, and era. In essence, form in music refers to the perceived relationship between events in time, including the arrangement of phrases in a song, the different sections that make up the song, and the song's duration.

An in-depth discussion on music form goes beyond the scope of this book. However, if we consider form in Western art music, many interpretations are centred around the connection between form and structure. The distinction between form and structure in music was not immediately apparent; I once saw them as the same concept. The form of a piece was merely its sectional order. However, this is just one way to perceive form, which can be seen as a mould or a predefined shape into which music is composed.

Fixed forms are often used as a tool for emulating different composers by providing a template for the arrangement of movements and themes in the work. A prime example is the Sonata form, a concept proposed by Adolf Bernhard Marx in 1845, which he derived from his extensive analysis of the works of Classical composers Mozart, Beethoven, and Hayden. Notably, this form or construct did not exist during the time the compositions were originally written; Marx formulated it based on the composers' practices that, in turn, created their musical forms (Burnham, 2002). It must be said that while the forms found in Classical or popular music are fixed, their interpretation by different audiences and generations is not fixed. This combined with their quintessential representation of a particular time period is why a work is called a classic.

When composing static music in popular contemporary styles, form is highly relatable in this context. Popular music forms entail the organisation of different sections, and these structures could consist of verses, choruses, bridges, or builds, drops, and hooks in electronic music. It is easy to view form as merely a mould or a structure into which you fit your composition when composing music of certain styles and genres. However, there are many other ways to interpret form, which I aim to elucidate.

It is an interesting question: did Mozart, Beethoven, and Haydn write music to fit a form that did not exist during their time? Or did they simply follow some sort of structure or rules that they believed would help them create the music they wanted to write?

Music from the Classical period (1750–1830) is typically divided into sections, generally determined by movements or key areas within a total framework. Popular music also has distinct forms, such as the 12-bar blues, song and strophic forms. However, an exciting aspect of form worth delving into in detail is the development of a unique form for each work in 20th-century Western art music, the Modernist period.

As 19th-century Western art music witnessed the gradual decline of tonality and increased use of chromaticism to the point of harmonic ambiguity, approaches to form during the late Romantic period (1830–1900) became less predictable. In response to this ambiguity, some composers, such as Brahms, utilised neoclassical forms to create large-scale works. Conversely, despite using neoclassical forms in his 12-tone system, Schoenberg, in the early 20th century, saw form as being evolutionary. 'Form means that a piece is organised; i.e. that it consists of elements functioning like those of a living organism' (Schoenberg et al., 1967).

Moving into the 20th century post WWII, with the emergence of experimental, avant-garde, and computer music compositions, many musical forms resulted from processes generated by algorithms, procedures, chance, directions, and randomness. This fresh approach to form, much of which does not observe functional tonality, is a defining force for the foundations of mid-20th-century experimental art music.

Distinguishing between popular music forms and Schoenberg's organic concept of form uncovers stark contrasts. Popular forms, such as verse-bridge-chorus, represent a static or fixed structure that poses challenges when creating Transmutable Music. To offer a different perspective on form, I intend to examine the ideas of three composers: Edgard Varèse (1883–1965), James Tenney (1934–2006), and Curtis Roads (born 1951). Their views on the relationship between form and structure will be considered.

Starting with Varèse. He does not perceive form as a template to be followed or a mould to be filled. To Varèse, 'form is a result - the result of a process' (Varèse, 1967). Each of his works, as he says, 'discovers its own form' (Varèse, 1967). To illuminate his understanding of musical form as a resultant, he draws an analogy between the formation of his compositions and the phenomenon of crystallisation.

I will attempt to summarise Varese's illuminating analogy. Think of the process of creating music as growing a crystal. Crystals are not made by pouring a substance into a predefined mould; they naturally take shape over time due to specific conditions and processes. Similarly, Edgard Varèse sees musical form not as a predefined structure to be filled in but as something that naturally evolves and comes to life during the creative process. In his words, 'Each one of my works discovers its own form' (Varèse, 1967).

Now, consider what makes a crystal. On the smallest level, you have individual atoms. These atoms interact with each other, joining together and growing over time. The crystal begins to take its unique shape as they join and grow.

This is like the different elements of a music composition – the melodies, rhythms, harmonies, and so on. As a composer, you have these individual elements at your disposal. As you arrange and combine these elements, they start interacting. Just like atoms in a crystal, they grow together and form the shape of your musical piece (Varèse, 1967).

In the case of a crystal, its final form is not predetermined. It is the result of how its atoms interact and arrange themselves. The same goes for a piece of music. Its final form is not a predefined structure but rather the outcome of how its different elements have been arranged and interact with each other (Varèse, 1967).

To put it simply, the inside structure of a crystal does not change much. It is the same across different crystals. However, the outer shapes the crystal can take, which form from these internal structures, are endless. Each one is unique because the internal parts combine in different ways (Varèse, 1967).

So, in Varèse's view, the form of a musical piece is a natural outcome of the composition process, much like the form of a crystal is a natural outcome of its growth process. Varèse's music consists of interactions of identifiable units. These units could be a rhythmic pattern, a collection of notes, a texture consisting of layers or fused elements. These continually interact with each other to form new sounds and combinations.

Varèse's idea that the resultant process involves organising sounds and sections in time and space echoes Schoenberg's evolutionary form. Similar to Schoenberg, Varèse uses these definitions of form to explain his process within non-tonal music, which in many instances are sonic entities and music centred on rhythm or timbre.

Let us now look at James Tenney's thoughts on musical form. His works, 'META/Hodos' (1961) and 'META META/Hodos' (1977), are significant. He uses the idea of a 'temporal gestalt' to describe how there are different layers within music and how those layers interact.

Imagine a song. We could break it down into sections, like the verse, chorus, and bridge. Within those sections, there are phrases. Within those phrases, there are individual notes or sounds. Each note or sound can be further broken down into elements like timbre and waveform. These layers are related, from the smallest (like an individual note's waveform) to the largest (like the overall song structure).

Tenney uses the idea of a 'gestalt' to describe this. A gestalt is a pattern or structure that we perceive as a whole. So, in music, the 'gestalt' is how all the different layers and elements come together to form a complete song.

According to Tenney, the form of a piece of music is like its blueprint or framework. However, there is more to it than just how the sections of a song are arranged. He talks about three different types of structure: 'statistical structure,' 'morphological structure,' and 'cascaded structure.'

Statistical structure describes how elements vary within a specific range. For instance, if you had a sound changing in frequency, the statistical structure would describe how much that frequency varies (state).

Morphological structure refers to the shape or contour of the music over time. It is like the 'envelope' that wraps around each musical element or part (shape).

Cascaded structure refers to how the smaller parts fit together to make up the whole (structure) (Tenney et al., 1986).

So, in Tenney's view, form is about how these different structures and elements fit together at different layers or levels, just like how the structure of a crystal creates its overall form. Each level contributes to the total form, from the smallest to the largest. Moreover, this is why understanding form is crucial for making music that can change and adapt with data.

James Tenney's theory outlines a hierarchy of temporal levels in music, from smallest to largest:

- Sound Elements: The smallest, indivisible perceptual units of sound.
- Clangs: Successions of two or more sound elements.
- Sequences: A succession of two or more Clangs, creating a larger gestalt.
- Higher-Level Temporal Gestalts: These are larger musical structures, potentially corresponding to traditional concepts like 'movement' or 'section' (Tenney et al., 1986).

At each hierarchical level, there are also three types of structures to consider. These are the statistical structure (state), morphological structure (shape) and cascading structure or internal relations (structure).

Let us consider Tenney's theory using a metaphor of building a house:

Sound Elements: These are the bricks of our house. Individually, they might not mean much, but together, they form the fundamental building blocks of our structure.

Clangs: Think of Clangs as the walls we construct with our bricks (Sound Elements). Each wall is a unique combination of bricks arranged in a specific order. This could refer to a unique melody or rhythm pattern in a musical context.

Sequences: Like we arrange several walls to form rooms in our house, we arrange Clangs to form sequences. A Sequence might be a verse in a song where several melodic or rhythmic ideas (Clangs) are connected.

Higher-Level Temporal Gestalts: These are the complete sections of our house, like the kitchen, living room, etc. In music, this could refer to a complete verse or chorus section or even an entire movement in a symphony. These larger structures comprise a series of Sequences, Clangs, and Sound Elements.

In addition to these levels, we need to consider three types of structures at each level:

Statistical structure (state): Imagine looking at the house from a distance. You cannot see the individual bricks (Sound Elements) or walls (Clangs). Instead, you get an overview or general 'feel' of the house – its overall colour, the number of floors, the type of roofing, etc. This is the statistical structure – the broad overview, the music's overall 'state' or 'feel.'

Morphological structure (shape): This is like examining the unique architectural features of the house – the placement of windows, the curve of the archways, the design of the front door, etc. In music, this would refer to the unique 'shape' (or design?) or form of a melody or rhythm – how it evolves and changes over time.

Cascading structure or internal relations (structure): This is like understanding how the house's different parts connect. How does the kitchen connect to the dining room? How does the staircase lead to the upper floors? In a song, this would refer to how the parts (like the verse, chorus, and bridge) are connected and transition into each other.

Taking a step back and looking at the state, shape, and structure of each hierarchical level of music is the first step in beginning to compose Transmutable Music.

Advancements in electronic music composition, production tools, coding, and software now pave the way for a broader array of hierarchical layers, or gestalts, within music and sound. Building upon Tenney's concept of the gestalt, Roads, in 2001, devised a framework encapsulating a broader spectrum of hierarchical levels. This is based on the physics of music and sound. While this framework might seem extreme for most composers, it caters to the varied range of Transmutable Music forms, which may rely on highly complex algorithmic approaches.

Here are Roads' timescales, listed from the largest to the smallest:

- **'Infinite:** Ideal time span of mathematical durations, such as the infinite sine waves of classical Fourier analysis.
- **Supra:** Timescale exceeding that of an individual composition, extending into months, years, decades, and centuries.
- **Macro:** Timescale of overall musical architecture or form, measured in minutes or hours, or in extreme cases, days.
- **Meso:** Divisions of form. Groupings of sound objects into phrase structures hierarchies of varying sizes, measured in minutes or seconds.
- **Sound object:** Basic unit of musical structure, generalising the traditional concept of a note to include complex and mutating sound events on a timescale ranging from a fraction of a second to several seconds.
- **Micro:** Sound particles on a timescale extending down to the threshold of auditory perception (measured in milliseconds).
- **Sample:** Atomic level of digital audio systems: individual binary samples or numerical amplitude values, succeeding each other at a fixed time interval. The period between samples is measured in microseconds.
- **Subsample:** Fluctuations on a timescale too brief to be properly recorded or perceived, measured in nanoseconds or less.

• **Infinitesimal:** Ideal time span of mathematical durations, such as the infinitely brief delta functions' (Roads, 2001, 3).

Composing at some of these extreme hierarchical structures might be hard to envisage. However, it might become feasible as AI increasingly intertwines with our creative processes. Exploring hierarchical gestalt can be a captivating venture.

Jorge Luis Borges (1899–1986), an Argentinian author, provides a perfect analogy for understanding the infinite potential of micro to macro timescales. He says, 'The line is made up of an infinite number of points; the plane of an infinite number of lines; the volume of an infinite number of planes; the hypervolume of an infinite number of volumes' (Borges et al., 1977).

Jorge Luis Borges' analogy unravels a profound perspective on the depths one could traverse while creating Transmutable Music. It captures the limitless potential for exploration, from the macro down to the micro, and reveals the complex interconnections that exist within the entirety of the gestalt.

As a composer, embracing this approach is unusual but has its merits. It challenges the conventional limits of composition, stretching beyond the mere arrangement of notes within phrases, sections, rhythms, fields, styles, sound design and mixing. Instead, it nudges you to investigate the myriad ways a composition can transform and evolve.

The more detailed your exploration within the hierarchical structure, the more nuanced and intricate your composition becomes. It becomes more challenging as you dive deeper into this realm, but the potential for dynamic, adaptive, and deeply textured music increases exponentially.

This philosophical thinking not only invites us to comprehend the complexities and intricacies that make up the whole gestalt but also stimulates us to keep pushing boundaries, experimenting, and transforming the very art of music composition. Thus, the process becomes as transformative as the music itself.

Roads (2001) discusses approaches to composition at a macro level as either top-down or bottom-up. He explains, 'A strict top-down approach considers macro structure as a preconceived global plan or a template whose details are filled in later stages of the composition' (Roads, 2001, 12). Conversely, 'A strict bottom-up approach conceives the form as a result of a process of internal development provoked by interactions on lower levels of musical structure' (Roads, 2001, 13). To summarise, a top-down approach moulds the music to fit a specific form, whereas a bottom-up approach aligns with Varèse's description of form as a process's result.

Laurie Spiegel, a renowned composer and music technologist, stated in 2000, 'Much of the new music we hear today is not organised along hierarchical lines at all. It may be built by progressive layering of unrelated elements or extrapolation or permutation' (Spiegel, 2000, 151). This is what experimental musicians call process; the interest is in the way things are put together rather than the outcome. She adds, 'The likes of such standard multi-level forms as the

Sonata and Rondo are no longer used nor are compatible with new forms being invented.' The likes Nonetheless, while comparable new music forms have yet to be invented, new standard processes are (Spiegel, 2000, 152).

The concept of a note in music has transformed from being a necessity to an option in music creation. This shift allows us to construct musical movements within a single complex sound, structure, or texture by varying amplitudes and harmonic spectra. This approach to music making is becoming increasingly common (Spiegel, 2000, 152).

To summarise the discussion about form and structure as they relate to Transmutable Music, we can consider the structure of a piece as a product of variability within hierarchical levels, which can be influenced by data ranging from the infinitesimal to the infinite. Any level of this hierarchy can change in Transmutable Music. As a result, musical form is no longer fixed in time but is subject to reinvention, reinterpretation, variation, and disruption on either a micro or macro scale.

This new perspective poses challenges, particularly for composers who work within the confines of popular or traditional music forms that usually rely on repeating structures such as AB sections, verses, bridges, and choruses. Spiegel's perspective introduces the concept of process into composition, which may offer a solution to this challenge. The exploration of music compositions based on process is a topic we will investigate next.

Process and Musical Systems

Process-driven composition offers an array of tools and techniques for those intrigued by Transmutable Music. The term 'process' refers to a series of actions and steps that progress naturally or systematically towards a specific result. Nyman (2000) discusses various approaches to process in the composition of experimental music, which is not bound by predetermined or fixed forms (Nyman, 2000, 6–9). This perspective aligns with Varèse's definition of form, suggesting that experimental composers were more invested in the process than in the form.

This approach distinguishes static compositions (like a traditional song) from Transmutable Music, the latter of which is not confined to common song forms. When composing, the artist must decide on the process for organising their musical material. With popular music, this involves arranging and developing musical themes, harmonies, rhythms, song structures, and melodic structures. This type of music composition occurs on a timeline or a fixed sequence of events with a specific duration, such as a three-minute pop song. The work is fine-tuned to incorporate emotional dynamics, continuity, and an overall music form structure.

In contrast, when creating transmutable works, the composer uses a process to dictate the functionality of the music within the musical design. This process establishes a system or set of rules or instructions dictating ways the musical material can be organised in real-time. Instead of being composed on a fixed timeline, the work is built on processes, events, and transitions that can be ordered or occur in various ways, depending on the system driving the music.

Nyman (2000) describes categories of process in composition. These include chance determination, people, repetition, electronic, and context processes (Nyman, 2000, 6–9). Many of these processes have been extended and explored since then with generative, algorithmic and computer music techniques in system building.

What is fascinating is how experimental composers from the mid-20th century developed and explored tools and techniques for composing music, laying a foundational framework for music incorporating interactive technologies and beyond. Their approach is innovative and prophetic in its anticipation of the evolution of music.

The process-driven composition techniques and the emergence of computing have led to the development of system building in music, an idea that might seem baffling to those outside and even to some within the music field. The proposition of composing a system that generates music, rather than a singular, static composition like a three-minute pop song or three-part movement, represents a profound shift in perspective. Here, the focus is something other than what might sound 'better' to individual listeners. This is because musical perception is profoundly subjective and lies beyond our control.

In fact, the debate between process and product is a significant topic in academic literature. As Collins and Brown (2009) put it, 'the balance between systems building and final result is of great interest' (Collins and Brown, 2009, 2–3). In this context, it is important to acknowledge that the creation of a musical system does not equate to a relinquishment of creative control. Instead, it entails defining the rules that govern how the musical work will be played back, stipulating what can and cannot change. This shift in perspective underscores the creative control and intentionality inherent in designing a musical system despite the mutable nature of the resulting music.

Indeed, the concept of relinquishing creative control to a system that produces mutable, generative, or interactive music can baffle many. It is not uncommon for composers in this field to be questioned about giving away their creative control. However, it is essential to understand that when one creates a system, they are indeed composing a song or work. They establish the rules determining how the song or work will be played back, defining what can change and what cannot. The processes they employ to compose their work are the ones they have organised.

Music, after all, is organised sound, and are not all composers employing processes to structure their works, which in turn produce their unique forms? This perspective underscores the creative control and intentionality that goes into designing a musical system, even if the final product has the capacity to change.

I will explore these concepts in more detail in Chapter 6 on Transmutability. We will look at the relationship between music and data, musical representation models, choice and control, and an overview of the history of autonomous, random, stochastic music, algorithmic works, and systems.

Time and Music

Understanding music is like embarking on a journey through time, linking the present, past, and future. It relies on our ability to connect present musical moments with past experiences while anticipating what is to come. The heart of musical cognition lies in the suspense and fulfilment, or disappointment, of these anticipations, sparking a cycle of tension and release that forms the essence of many musical pieces (Kirby, 2016).

Music is intrinsically time-based. A singular work generates its unique temporal space and unfolds within it. Put more simplistically, it creates its own time while also being experienced in time. Given its potential for change at any instant, time becomes a critical issue in the production and experience of Transmutable Music. Here, I introduce two concepts of time: the 'timeline' and the 'time stream.' A timeline is a sequence of events or points that follow a logical order, such as past, present, and future. This linearity often characterises music, depicted in musical notations, software environments, and gaming compositions, using analogies like horizontal and vertical changes. Borrowed from science fiction, a time stream (Kaae, 2008, 76) metaphorically represents a non-linear sequence of events, akin to filmic flashbacks or future flashes, occurring in any order.

Fixed music forms are structured around a timeline, where events transpire in a definitive order across all renditions. This timeline is linear, arranging the music to induce a particular emotional or conceptual impact. Most important is that a fixed musical form can be repeated. It embodies a structure composed for a start-to-finish experience over a specified duration.

In contrast, a time stream is a scenario where events can materialise in any order, depending on certain conditions. These events are not shackled to a specific timeline. The works of Stockhausen, such as his *Momente* (1962–68), John Cage, and Earl Brown illustrate such musical time streams. Transmutable Music, thus, can be understood as residing within a time stream. It does not confine all musical aspects to a timeline. Instead, it relies on a dynamic approach to time, manifesting as events and processes occurring within a time stream, responsive to data fluctuations. The set of rules or a control system, designed by the composer, allows events triggered by people, machines, or external factors to interact with others. Hence, Transmutable Music allows composers to traverse the manifold dimensions of time.

While Transmutable and static music forms intersect at many levels, their representation diverges significantly. Fixed music forms, represented via notation and charts, contrast with the visual representation challenges posed by Transmutable Music. The latter's ever-evolving, open form or control via a control system makes visual depiction difficult. In Transmutable Music, the work's form continually takes shape through composer-designed processes. Nevertheless, a system is indispensable for the experience of Transmutable Music.

Transmutable Music System

In our whirlwind journey through music's various elements – form, structure, time, and process – we have seen how they come together within a system. Just as traditional music requires a playback device, such as a CD player for CDs or a streaming platform for MP3s, Transmutable Music demands a system tailored to its dynamic nature.

In Chapter 3, we will look at Transmutable Music Systems. To create this type of music, creators need to design a system that uses data inputs to modify musical parameters, resulting in the music that listeners hear. Although it may seem complicated, do not worry; this book will explain the components and workings of a Transmutable Music System in an easy-to-understand and comprehensive way.

A Transmutable Music System is multifaceted. It can be designed alongside the music, enabling the live creation of the work's content and form. Alternatively, it might be applied to pre-existing compositions, offering a range of musical choices that the system brings to life through interactive playback. Such systems can also be developed to generate musical design in real-time, like creating synthesisers or organising sound materials.

These systems vary, but whether they are interactive, adaptive, reactive, contextual, autonomous, generative, or powered by AI and algorithms, they share a fundamental process. They map data inputs to modify the music's state, structure, or parameters at different levels, resulting in the 'rendered musical output.'

To contextualise the Transmutable Music System within a practical setting, consider the workflow depicted in Figure 2.1. Here is a brief overview of the elements within a Transmutable Music System:

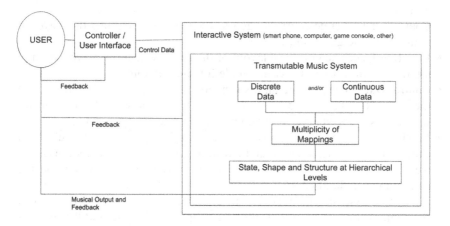

Figure 2.1 Transmutable Music System Diagram.

- **User:** This could be a listener or performer interacting with the system.
- **Controller:** A device or interface, like a touchscreen or joystick, through which the user engages with the music.
- **User Interface:** The visual or tactile interface that facilitates user-system interaction.
- **Control Data or Data:** The input from the user or contextual sources that the system will process and map to musical parameters. This can be discrete or continuous data, which is discussed in detail in Chapter 3 in the Control System section.
- **Feedback:** The system's responses to user actions, tactile or visual, help the user understand their influence on the music.
- **Interactive System:** The hardware or software platform that plays back the music, such as a smartphone, computer, or custom-built system.
- **Rendered Musical Output:** The final music output, shaped by the user's interactions with the system.

Video Game Example

In a music-based video game, the user – the player – engages with the game through a controller, utilising buttons and a joystick to influence the gameplay. Each interaction generates control data, which is transmitted to the game console. The game operates on a platform like Unreal Engine or Unity, where audio middleware is incorporated. This combination of the game engine and the audio middleware constitutes what is known as the Transmutable Music System.

This system's control data directly influences the music and sound effects played, producing the final audio-visual experience. The game provides visual and audio feedback so the player recognises that their actions have direct consequences within the game environment.

- **User:** In this case, the game's player interacts with the system as a listener or a performer.
- **Controller:** The game controller acts as the interface, enabling the player to manipulate the game and its music.
- **User Interface:** The graphical layout of the game, which the player sees and interacts with, provides a visual guide for their actions.
- **Control Data or Data:** The inputs generated from the player's actions with the controller. This could be pressing a button or a joystick's movement, which the system will interpret and map to change the music and sound FX.
- **Feedback:** The game's response to the player's inputs, which could be a change in the music, a new sound effect, or a visual cue on the screen. This feedback helps the player understand the impact of their actions on the game's musical landscape.
- **Interactive System:** This includes the game console or computer on which the game is running, as well as the game engine and audio middleware that work together to process the control data and generate the music.

- **Rendered Musical Output:** The music and sound effects the player hears, shaped and altered by their interactions with the game, create a unique musical experience each time they play.

Interactive Mobile App Example

In an interactive music mobile app, the user interacts with the app through the touchscreen of their mobile device. Taps, swipes, and other gestures are all forms of input that generate control data, which is then processed by the app. The app, powered by its underlying software, acts as the Transmutable Music System.

As the user interacts with the app, their gestures determine how the music is played and modified, resulting in a dynamic audio experience that can change with every interaction. The app offers immediate visual and auditory feedback, ensuring that users understand the relationship between their actions and the changes in the music.

- User: The individual using the mobile app can be both a listener and an active participant in shaping the music.
- Controller: The mobile device's touchscreen, through which the user directly interacts with the music using touch-based gestures.
- User Interface: Uses a graphical user interface (GUI). The app's design and layout are displayed on the screen, which guides the user's interactions and provides visual feedback to their inputs.
- Control Data or Data: The data generated from the user's gestures on the touchscreen. Whether the user is tapping to the beat, swiping to change tracks, or pinching to alter effects, the app collects and interprets this data to affect the music.
- Feedback: The visual animations or changes in the app, alongside the auditory changes in music, inform the user about the results of their interactions.
- Interactive System: The smartphone or tablet with the app installed includes the software necessary to process user inputs and generate musical changes.
- Rendered Musical Output: The altered and interactive music that the user hears directly results from their interaction with the app. This output is not static but is continually reshaped by the user's real-time engagement with the app, creating a personalised musical journey.

A Transmutable Music System integrates these elements to allow for a music experience shaped by user input, transforming listeners into active participants in the music creation process.

Summary

This chapter explored Transmutable Music, a novel and evolving musical form that provides dynamic and continually changing sonic experiences.

Unlike conventional, static forms, Transmutable Music adapts with data. By establishing an overarching term, I aim to formalise and unify approaches to designing music that can adapt and transform in response to data. This framework is developed to simplify the intricacies for composers and producers working in this field.

We examined the ideas and approaches of five eminent composers – Edgard Varèse (1883–1965), James Tenney (1934–2006), Brian Eno (1948), Laurie Spiegel (1945), and Curtis Roads (1951) – to gain insights into the form, structure, and time in Transmutable Music. Form was defined as the organisation or generation of a musical composition and its relationship to time. Varèse posited form as the outcome of a process rather than a mould to fill (Varèse, 1967).

Hierarchical levels in music and the ability to manipulate shape, state, and structure at these levels unveil the potential for data to transform musical form. Tenney's three primary forms of structure – statistical, morphological, and cascaded structure – offer a fundamental understanding necessary for algorithmic, deterministic, and stochastic music (Tenney, 1986). Roads (2001) extended Tenney's levels and provided nine timescales based on the physics of music and sound.

Transmutable Music also challenges traditional perceptions of time, building on temporal processes introduced by experimental music and the Avant-garde of the mid-20th century. We introduced the metaphor of a 'time stream' in contrast to a 'timeline' for fixed forms of music.

Transmutable Music requires a system for mapping data to musical parameters and rendering musical output. The following chapter will discuss how to compose Transmutable Music and its system, which will explore the musical design and essential components of Transmutable Music in detail.

References

Borges, Jorge Luis, Norman Thomas Di Giovanni, and Alastair Reid. 1977. *The Book of Sand*. New York: Dutton.

Burnham, Scott. 2002. "Form." In *The Cambridge History of Western Music Theory*, edited by Thomas Christensen, 880–906. The Cambridge History of Music. Cambridge: Cambridge University Press. 10.1017/CHOL9780521623711.030.

Collins, Nick, and Andrew R. Brown. 2009. "Generative Music Editorial." *Contemporary Music Review* 28 (1): 1–4. 10.1080/07494460802663967.

Etymonline. 2023. "'transmute' Etymonline.'" Etymonline. 2023. https://www.etymonline.com/word/transmute

IFPI. 2016. "Global Music Report." http://www.ifpi.org/downloads/GMR2016.pdf.

Kaae, Jesper. 2008. "Theoretical Approaches to Composing Dynamic Music for Video Games." In *From Pac-Man to Pop Music: Interactive Audio in Games and New Media (Ashgate Popular and Folk Music Series)*, edited by Collins Karen, 75–91. England: Ashgate Publishing Company.

Kirby, F. E. 2016. "Musical Form." Encyclopædia Britannica, Inc. 2016. https://www.britannica.com/art/musical-form.

Nyman, Michael. 2000. *Experimental Music Cage and Beyond*. 2nd Edition. United Kingdom: The University Press, Cambridge.

Roads, Curtis. 2001. *Microsound*. Cambridge, MA: MIT Press. http://ebookcentral. proquest.com/lib/newcastle/detail.action?docID=3339387.

Schoenberg, Arnold, Leonard Stein, and Gerald Strang. 1967. *Fundamentals of Musical Composition*. London, United Kingdom: Faber & Faber.

Spiegel, Laurie. 2000. "Music as Mirror of Mind." *Organised Sound* 4 (3): 151–152. 10.1017/S1355771800003046.

Tenney, James. 1986. *Meta + Hodos: A Phenomenology of 20th Century Musical Materials and an Approach to the Study of Form and Meta Meta + Hodos*. Oakland, CA: Frog Peak Music.

Varèse, Edgard. 1967. "Rhythm, Form and Content." *Contemporary Composers on Contemporary Music*. New York, NY: Norton.

3   Components of Transmutable Music

Components of Transmutable Music

We are now getting to the heart of the book. We will apply the main ideas discussed in the early part of this section – form, structure, process, time – as well as the role of Transmutable Music within a system. If you jumped to this chapter to delve into the compositional model, I strongly recommend revisiting the first chapter of this section. It establishes a foundational understanding, which is necessary for what follows.

Our discussion is backed by various research fields – electronic music creation, interactive systems, generative and algorithmic music composition, music theory, and game audio. Composing Transmutable Music is not about creating a static song like traditional music. Instead, it is akin to designing a 'song system.'

The Merriam-Webster dictionary defines a 'system' as a group of interrelated parts working together, much like machine components or network nodes. It is a complex whole, or alternatively, a set of protocols or procedures governing a specific process.[1]

Transmutable Music involves numerous system types, which can create, change, organise, and playback music.

We will explore four primary parts of a Transmutable Music system here:

Content – This is the material utilised within the system. It can encompass sound, music, code, images, videos, and animations.

Musical Architecture – This is the form and structure of the music piece. This is where the content is shaped and structured.

Control System – This is essentially a system that oversees and regulates another system. It where data is organised and mapped to control the content and musical architecture.

Experience – Music, 'experience' encapsulates the relationship between the composer, the performer, and the listener.

Figure 3.1 visually represents the interactions among all components of a Transmutable Music system. This section will explore these four components,

DOI: 10.4324/9781003273554-4

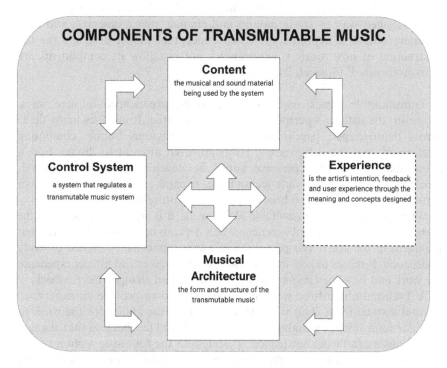

Figure 3.1 The Components of a Transmutable System.

examining their interrelationships and implications for designing a Transmutable Music System.

Designing Experience in Transmutable Music

Experience within a Transmutable Music work is a complicated beast. It can be understood from a composer's and listener's perspective. The listener's perspective is a large and complex area involving cognition, perception and emotional states. Spiegel (2000) states:

> As more is learned about the way we ingest, parse, and experience sounds, the technical rubric on which our creative concepts take form would be designed for optimal plug-in compatibility with our music's intended destination, the receiving processor, the ear and mind. (Spiegel, 2000, 152).

Spiegel (2000) highlights the potential for future music to be composed with an individual's perception in mind. She states that instead of music represented in models that explain its structure and components, there could be a conceptual model for how music is experienced. This model could become as important as the compositional processes used.

'As calculus introduced to mathematics in the 17th century a previously unimagined way to study motion and change, music deserves, and may finally obtain, a representation and conceptual model designed for the structure of how music is experienced, not just how its components are manufactured' (Spiegel, 2000, 152).

Transmutable Music requires music to be experienced in new ways. However, the listener's perspective is a complex area. It involves many fields across neuroscience, psychology, cognitive behaviour, human computing interaction, and the list goes on. I will not even attempt to discuss how a conceptual model of experience could be developed. However, as AI, blockchain, and personal data technologies expand, models based on human experience are predicted to become important future commodities.

When producing a Transmutable Music work, it is crucial to understand the design of the desired musical experience. This depends on the overall concept and the intent of the artist. The perception of music cannot be controlled or easily understood. It relates to how the audience, performer, user and listener experience the work and how they might understand the musical changes and feedback.

A Transmutable Music work must be designed to provide an experience beyond a static form. This will include designing the ability for the work to produce some level of variability in its playback and feedback so that changes in the music can be acknowledged. Throughout the following section of this book, we will look at a palette of approaches available for artists to create these new immersive experiences.

Creating Experience in Transmutable Music

Due to the vast options available for designing experience in Transmutable Music, I have summarised some considerations you may want to consider when exploring your initial ideas of creating a Transmutable Music work.

1 Listener Engagement and Experience

- The duration of time the listener is expected to engage with the work.
- The number of times the listener is expected to want to experience the work, a one-off experience or repeated listening.
- The number of people using or listening to the work simultaneously: multiple users or listeners.
- The other media available to provide feedback about changes in the music, visuals, haptics.
- The other media involved in the work, e.g. audio-visual work.
- Does the music need to be at the forefront of the experience, or does it play a more supporting role?

2 Narrative and Experience Design

- The time it takes to experience the story or narrative.

- The amount of variability the composition might need to keep the listener engaged.
- Does the music transition smoothly to new parts or sections?

3 Interactivity and Control

- How will the music change and adapt, and how much control will the listener have over the music?
- Game-like or goal-oriented interaction design.
- How can you ensure the user understands how to interact or experience the work?

4 Creative Direction and Composition

- How much music needs to be composed based on the length of experience?
- Creative direction and musical style/s.
- How noticeable will the musical changes be?

5 Technical and System Considerations

- System limitations and requirements, for example, memory issues and content limitations.
- The content (music and sound material) that will be used, any digital instruments required, and production tools that can be used.

This list of considerations highlights design attributes for creating and refining experience in a Transmutable Music work. These considerations are just the beginning of refining and developing user experience for your work. However, you should never assume that you will understand how your listeners and users will understand your work. Some of these considerations may require user experience evaluation to help provide better experiences and assist the artist in making sure their work can be understood. Chapter 4 discusses types of user experience evaluation in detail. However, a good start would be to explore the types of feedback you can design for the listener, user or performer. Feedback is an excellent tool available to composers.

The Role of Multimodal Feedback in Enhancing Transmutable Music Engagement

Feedback activates engagement and a deeper understanding of the work. Feedback may include one or more of the following.

- Visual
- Auditory
- Tactile
- Kinesthetic

Designing feedback is an essential part of the experience design. Transmutable Music offers an experience beyond a static form. The listener, user, and performer must understand how their actions or data affect the musical output. Wanderley and Depalle (2004) examine four types of feedback based on the following characteristics;

• Primary feedback includes visual, auditory, and tactile or kinesthetic feedback.
• Secondary feedback is the sound produced by the instrument.
• Passive feedback is 'provided through physical characteristics of the system (a switch noise, for instance).'
• Active feedback is a response to ascertain user action. This is produced by the system' (Wanderley and Depalle, 2004).

Evaluating feedback from a system, as a whole, can be achieved through Music Interaction approaches. However, feedback also becomes part of the musical design. Producing Interactive (or Transmutable) Music requires designing all possible actions and interactions and their effect on the musical architecture. Producers must consider how they can keep their listeners and performers engaged and involved in the work from a musical standpoint. This also means to compose music that has obvious changes. Changing the lead vocal/melody or dramatically changing the tempo/genre are examples of obvious changes for audiences who may not notice more subtle musical changes.

Feedback can also guide a listener and engage them to take action. For example, the work could resolve to the tonic once the user performs an action. If the work involves multiple mediums, the feedback may also include visual and haptic experiences. A button flashes in colour, indicating it wants to be pressed. Every time the user presses that button, a slight vibration indicates that an action has been produced. Studies highlight the importance of providing feedback to individuals to create meaningful experiences and avoid frustration (Hodl et al., 2014; Lee and Freeman, 2013).

Designing the experience of a Transmutable Music work consists of several considerations, including feedback and control. Each work is designed based on the artist's vision, which should stipulate the amount of control required by the user, machine, performer, and audience. Chapter 4 will discuss experience design in more detail, including how to evaluate user experience.

The Musical Design for Transmutable Music

Artists use various approaches when composing Transmutable Music, depending on the specific form of Transmutable Music they are working with. A composer for generative music might construct a series of patches within Max or PD or use a coding tool like Chuck. Regardless of the software choice, all Transmutable Music necessitates a system. Crafting this system to create and change music requires a unique hierarchical compositional model.

The diversity of processes and control systems encompassed by the term Transmutable Music is vast. However, they can be understood within a compositional model. A compositional model is a composer's representation of the musical design for a piece. Italian composer Stefano Di Scipio articulates a compositional model as the composer's instantiation of a procedural description for the musical structure (Di Scipio, 1994). Building on this concept, producer and composer Greg White provides an excellent model of musical design that organises sound material into a musical architecture using control systems (White, 2015). He maintains that software environments facilitate real-time simultaneous performance, composition, and production. This convergence resonates deeply with Transmutable Music, as it is performed, composed, and mixed in real-time within its system.

White describes musical design as formulating a plan for music creation and the plan itself (White, 2015). Adapting White's model, I have developed an abstract model for the musical design of Transmutable Music, as illustrated in Figure 3.2. I have made a few adjustments, employing the term 'content' instead of 'sound material' to encompass a broader range of material that might be used. Transmutable Music often incorporates other media like visuals, video, text, animation, etc. This model should be employed for technical considerations when designing a compositional model for a Transmutable Music system.

Further explanation of this model lies in its role and purpose.

'Content' refers to various digital media types created, manipulated, and reconfigured within the work. This can be music, sound, animation, video, text, code, sound libraries, patches, procedural sounds, etc.

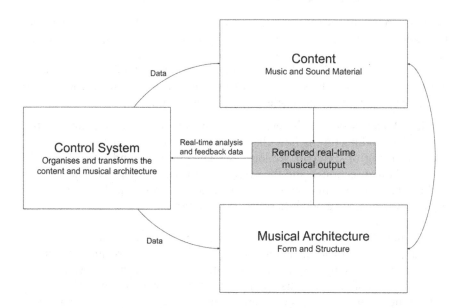

Figure 3.2 Components in the Musical Design Model for Transmutable Music.

'**Musical Architecture**' is the development and invention of music and sound at each hierarchical structural level.

The '**Control system**' organises, transforms, synthesises, programs, and analyses musical material and architecture into its rendered musical output.

Experience is implicit in this model's design within the control system, the content, and the musical architecture. It represents the artist's intent, the feedback, and the user's experiences of the meanings and concepts that have been designed. It is challenging to illustrate 'experience' within this model because it encompasses all areas.

Indeed, this is a simplistic model, but it is essential to remember that simplicity does not necessarily imply a lack of complexity or depth. The model provides a broad, abstract structure in which the control system organises and transforms data to manipulate and structure the sound material into the work's form and structure or the musical architecture.

Let us explore the three primary components of this model in more detail – the control system, the content, and the musical architecture:

Control System – This is essentially the 'brain' of the musical piece, organising, transforming, and synthesising the data to influence the other elements of the model. The control system can include software like DAWs (Digital Audio Workstations) and Max, generative algorithms, AI-based tools, or physical instruments and devices. The control system's complexity can range from a simple set of pre-programmed instructions to a dynamic, adaptive AI system that changes the music in response to various inputs or conditions.

Content – refers to the raw material or 'ingredients' used to create the music. In traditional music, this might be the notes, loops, samples or stems. However, in the context of Transmutable Music, this could extend to sound libaries, patches, MIDI, algorithmically generated sounds, or even non-musical data that can be sonified or used to influence the music. The control system can manipulate the **content** in countless ways.

Musical Architecture – This refers to the overall structure and form of the music. The architecture is how the content is arranged over time, including the sequencing of different sections, the layering of sounds, the dynamics (loud and quiet sections), and the overall 'narrative' of the piece. The musical architecture is also influenced by the control system but typically according to higher-level rules or guidelines set by the composer.

The interplay of these three components can create an infinite variety of musical pieces, each with its unique characteristics and complexities. By understanding and manipulating each component, a composer can use this model to create Transmutable Music that adapts and changes with data in real time.

I will now unpack these three components further to show the potential of Transmutable Musical design.

Content

The content component in this musical design model pertains to the media or material utilised in the system. It falls into two categories: composing with pre-existing material or generating material via the control system (or data). The material includes music, sound, images, videos, visuals, animation, etc. However, since this book focuses on music, the discussion will concentrate on music and sound materials (or content).

The control system selects, organises, transforms, creates, shapes, and integrates the content in a Transmutable Music system. A wide array of content can be modified and organised within a Transmutable Music system. However, it is crucial to remember that sound and music content often have their own inherent musical architecture. For instance, a loop or a stem recording would have its own structure and form.

The timescales discussed by Roads (2001) in Chapter 2 expand the musical design process to minuscule and even infinitesimal levels in the era of computer music, digitisation, and software environments (Roads, 2001). Given these extended design levels needed to create sound, musical architecture can be divided into two processes:

1 composing the sound, and
2 composing with sounds (Di Scipio, 1994).

Composing the sound refers to the creation of the sonic material itself, which electronic music processes like sampling and synthesis have greatly simplified. Composing with sounds involves using these created sounds, or traditional instruments, to compose higher-level structures, such as melodies, chords, and rhythms.

By separating these two processes, we can understand the complex role of music and sound content within a Transmutable Music system. However, sometimes, limitations are placed on the type of sound material that can be produced, particularly with mobile apps, to optimise memory and system requirements.

Music and sound content may consist of one or a hybrid of many sound and music creation approaches, including audio, MIDI, digital signal processing (DSP), digital instruments, plugins, patches, sound libraries, algorithms, procedural sound, and code. This could involve generative, stochastic, and algorithmic music and sound produced by AI. The data could be used to create the sound and/or organise the sound, similar to what occurs in electronic music production.

Organising the Content

In this section, various ways of organising content are discussed. Given the myriad of relationships across various hierarchical levels, organising content within a Transmutable Music system can be a complex endeavour. As simple as labelling audio or creating hierarchical filing systems may seem, techniques from game audio composition can be valuable in organising sound and music material to integrate into musical architecture.

Metaphors such as 'vertical' (indicating layering and blending) and 'horizontal' (signifying sequence of events such as song/formal structure) frequently illustrate how music is manifested on a timeline, similar to a musical manuscript (the instruments are presented vertically and the song structure unfolds on the horizontal) or the tracks used in DAW software. However, as underlined by various case studies within this book, new groupings and associations can be formulated beyond these vertical and horizontal metaphors. This is further discussed in Chapter 5. Transmutable Music presents an array of associations across multiple hierarchical levels, demonstrating the potential to be flexible, reminiscent of the infinite possibilities in a time stream.

The organisation of musical content can be approached in various ways. For example, it could be according to genre, song structure, formal frameworks, or characteristics such as tempo, frequency spectrum, onset, attack, time signatures, scales, etc. The *Ninja Jamm*[2] system serves as an example, employing music packs as its core content, offering users the choice of loops and enabling them to construct a remix. Users can also process sounds and loops across four tracks with distinct effects and digital sound processing capabilities.

The intricate nature of the system is a result of its usage of four groups of eight loops and a singular group of samples that can be activated within the app. These groups function as loop folders and are categorised as instruments such as drums, bass, synth, etc. The music packs are customised and curated by artists from the Ninja Tune label, with their titles and organisations hinging on how the artists utilise their content and materials. An example of such a filing system might be master/drums/loop1.

The hierarchical structure of the content organisation can be visualised through Figure 3.3, which demonstrates the various levels of the content. At the apex is the complete duration of the song generated on the app (Master Level), followed by levels for the instruments or tracks (Track Level), and subsequently, the loops, which are phrases of predefined lengths such as 8 bars (Phrase Level). These loops at each hierarchical level can be altered and transmuted using digital sound processing and customised devices/tools.

Thalmann et al. (2016a) introduce the Mobile Audio Ontology MAO[3] to define Dynamic Music Objects (DYMOs)[4] for more complex systems. This offers another way to organise and group relationships for Dynamic Music. It also visually provides a method which enables relationships which is hard to represent visually. (Thalmann et al., 2016b) (Figure 3.4).

The mobile ontology represents these relationships using a semantic web graph. Figure 3.4 represents a dynamic song, including all its relationships at different timescales. This image is presented in 2D; however, a 3D structure would be easier to understand. The outer layer of the diagram is where the music is played back, like a record (Thalmann et al., 2016b).

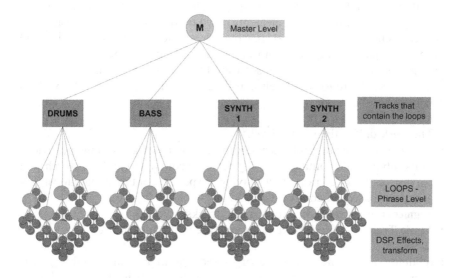

Figure 3.3 Ninja Jamm Content Organisation Schematic.

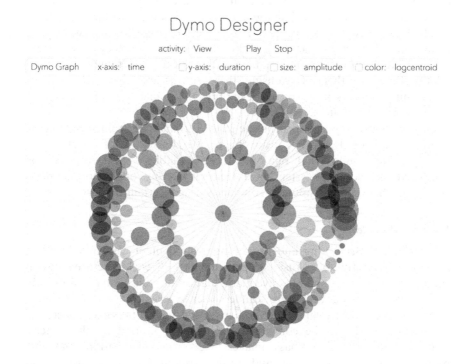

Figure 3.4 Example of a DYMO Reprinted from (Florian Thalmann et al., 2016a).

Ultimately, numerous ways exist to tailor a hierarchical filing system encompassing grouped audio elements based on the nature of the music being crafted (generative, contextual, reactive, or adaptive). This is a well-versed process for game composers who fabricate their custom musical and sound hierarchies in audio middleware software like *Wwise* or *Fmod*.

In *Wwise*, the following levels can be organised and designed by vertical and horizontal facets:

- The **work unit** is the master level.
- **Containers** can be multiple or nested. Each container dictates the behaviour of its subordinate containers or segments. A container can be a switch or a playlist, capable of randomly playing a specific container or choosing a container based on control data.
- **Segment**s organise and contain music tracks.
- **Music Tracks** incorporate and arrange the audio files in use.

Explaining how content can be organised within a Transmutable Music system can be likened to structuring a filing system. It is about creating an organisational framework to manage your content effectively according to the system's requirements. These 'filing systems' or hierarchies can incorporate a range of relationships and organisational data, thus allowing for more streamlined control and manipulation of content within the system.

Think of these systems as advanced folders or cabinets that store your content and manage how they relate to each other and how they are accessed. This is similar to creating a database with various data tables, each with relationships to others. Organising these tables and their relationships significantly impacts how you retrieve and manipulate the data, which, in this case, is your content.

Some systems may also incorporate groupings and constraints that trigger specific sounds or effects. For instance, you may have a list of variables and corresponding sounds that relate to those variables. This setup can be compared to coding a system in JavaScript or any other programming language, where specific inputs or conditions trigger certain outputs or events.

Summary of Content

In conclusion, structuring and organising your content can directly affect how your Transmutable Music system functions. A well-structured system enables more efficient and effective manipulation of your content, ultimately leading to a smoother and more seamless user experience.

Figure 3.5 encapsulates the content used in a musical system of Transmutable Music; the content is the material utilised by the system. This content may encompass images, video, and animation, but this particular model is for sound and music material, which includes digital instruments, audio, MIDI, plugins, patches, sound libraries, and procedural

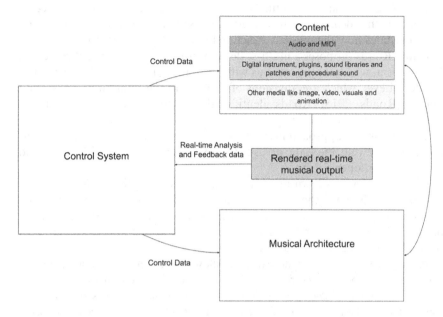

Figure 3.5 Musical Design of Transmutable Music – Content.

music. The organisation of the music and sound material hinges on the design of the control system and the musical architecture.

Musical Architecture

Designing the musical architecture for Transmutable Music, or in other words, crafting its form and structure, could be aptly described as 'composing with sound' (Di Scipio, 1994). This architecture encapsulates the musical or sound content and its careful arrangement, transformation, sequencing, balancing, and mixing to produce a final musical output.

The entire musical design is essentially a guide on how the control system manipulates and adjusts the content and structure on multiple hierarchical levels. This requires the composer to view the composition process in a new way. There are no strict rules – the design is entirely contingent on the composer's vision, the system design's constraints, and the interactive technologies in use.

These hierarchical levels of music can be understood by using the analogy of a puzzle that can be dissected into smaller pieces. For instance, a song can be split into sections like a verse or a chorus. These sections can be further divided into phrases or organised by layers, bars, etc. The hierarchical levels of musical form provide a foundation for composing Transmutable Music, where a system is developed to enable data to modify the musical architecture and content.

Designing musical architecture involves formulating a plan or a set of instructions that define how the content is utilised or created within the system. It also dictates how the music is structured into a larger or macro form.

Curtis Roads proposes a multi-scale composition consisting of three levels: Macro form, which is synonymous with the term master level; Substructures, or Meso Level, which are 'divisions of form'; and the sound material (sound object) (Roads, 2015). Roads identifies three basic approaches: top-down, bottom-up, and multi-scale.

The bottom-up approach involves designing from lower hierarchical levels upwards, allowing the data or user to create substructures with the sound material in an open form.

Conversely, a top-down approach implies designing the formal structure first and then filling in the lower levels to fit within this overarching structure. Top-down approaches are required when composing for a pre-existing system or control system, which requires content to be in a particular form. Like *Ninja Jamm*,[5] Nagual Sounds,[6] and *Weav*.[7]

A multi-scale approach is more flexible, allowing the composer to operate freely across timescale boundaries, re-evaluating and modifying the strategy at any stage. This is particularly beneficial for Transmutable Music, which is meant to be transformed and altered across all hierarchical levels by design. 'When composing for a timeline, many concepts are lost or not explored due to the limitations of the formal structure' (Redhead, 2020, 95). Transmutable Music works 'can include multiple sections, lyrical ideas or perspectives, changes in tempo,' feel or style. 'In fact, the musical design becomes boundless, giving composers the freedom to explore multiple themes and concepts in the same work (Redhead, 2020).

Macro Form

The macro form represents the overarching structure of a composition derived from a carefully designed compositional plan. Top-down is like designing music for a mould, where the composer starts with the design of a formal structure. The lower levels are then designed to fit within this mould. Top-down approaches are required when composing music for an existing Transmutable Music system, which has a pre-designed system.

Sub-structures

Substructures existing within the macro form can be motifs, phrases, sections or segments at the meso level. Roads (2001) identifies four common mesostructures:

'Repetitions – the most basic musical structure: iterations of a single sound or group of sounds. If the iteration is regular, it forms a pulse or loop.

Melodies – sequential strings of varying sound objects forming melodies, not just of pitch, but also of timbre, amplitude, duration, or spatial position.

Variations – iterations of figure groups under various transformations, so that subsequent iterations vary.

Polyphonies – parallel sequences, where the interplay between the sequences is either closely correlated (as in harmony), loosely correlated (as in counterpoint), or independent; the sequences can articulate pitch, timbre, amplitude, duration, or spatial position' (Roads, 2015, Ch 9, 33–34).

Sound Material (Sound Object Level)

The sound object level includes the fundamental unit of composition, the note, as well as any individual sound from any source.

A note, the key unit of conventional music architecture, consists of four properties: pitch timbre, dynamic marking, and duration (Roads, 2015, 13).

A sound object consists of various components such as pitch, amplitude, spatial position, and timbre. These components have multiple time-varying envelopes, which can change over timescales longer than conventional notes (Roads, 2015).

Developing a Transmutable Music system that has an open form with all content being at a sound object level could be confused with designing a digital musical instrument. This approach blurs the line between an instrument and Transmutable Music. It may be difficult for the user to construct sections and phrases, etc., to produce a work. However, this may be the design or the work that might work using a layered compositional approach involving adding and subtracting layers of sound.

Hierarchical Levels in Popular Music

In modern popular music making, the terminology is somewhat different. Terms such as master, duration, sections, phrases, loops, samples, soundbites, notes, builds, hooks, drops, breakdowns, and riffs can be introduced depending on the composer's intention.

Based on Road timescales as discussed here and in Chapter 2, the following hierarchical levels are suggested, see Table 3.1:

Master Level (macro level): This is the duration of the work and the overall form. It can be measured in minutes, hours, days, months or years, depending on the musical design. In a recording context, it is also the

Table 3.1 Time Scales Comparison

Macro Level	Master Level
Meso Level (substructures)	Section Level, Phrase Level
Sound Object	Sound Object

master mix level. Changes at this level affect all levels within' (Redhead, 2020, 96).

Section Level: A section of a track refers to one of the sections in a song structure, for example, verse (A), chorus (B), bridge, hook, breakdown, or drop. Sections consist of two or more phrases. A section can be measured in seconds or minutes (Redhead, 2020, 96).

Phrase Level: When two or more sound objects are combined to form an identifiable unit. It is also referred to as a motif. For example, a loop, an audio file, a small phrase of notes or a short melodic, harmonic or rhythmic motif. A phrase is measured in seconds (Redhead, 2020, 97).

Sound Object Level: This is a note or sound that might have pitch, duration, timbre, sound envelopes, waveform, amplitude, time, filters, and MIDI parameters.

Micro Level: This level involves 'sound particles on a timescale that extends down to the threshold of auditory perception' (Roads, 2001, 4).

Summary of Musical Architecture

The musical architecture in a Transmutable Music System is a map for composing with sound. It defines how the control system alters and transmutes the content and musical architecture on various hierarchical levels. The hierarchical levels, as showin in Figure 3.6, include the macro,

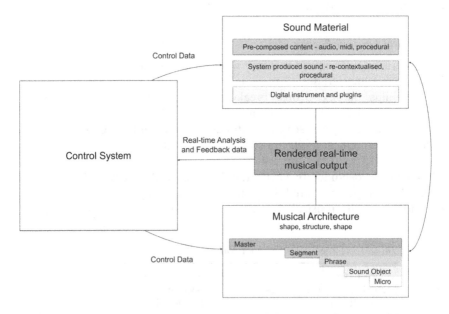

Figure 3.6 Musical Design of Transmutable Music – Musical Architecture Added.

meso, and sound object levels, which relate to the macro, substructures, and sound material in multi-scale composition.

Control System

Transmutable Music relies heavily on the control systems that regulate its playback and processes. Whether developed from scratch, adapted from existing software, or tailored for a specific musical project, the control system plays a pivotal role in transforming input data into Transmutable musical output. The following section provides a comprehensive overview of the role of control systems in Transmutable Music.

Control systems in Transmutable Music works can be:

• Part of the compositional process.
• Supportive tools that aid these processes.
• Frameworks that regulate the rendered music output.

Depending on the project, the control system can be:

• Custom-built specifically for the music work.
• Developed within a specific software environment, as seen in-game music composition.
• Predefined systems like the *Ninja Jamm* or the *Weave* app offer a template with the control system and its parameters already set.

When constructing a control system from scratch, several elements must be addressed:

• Type of data in use and its transformation method
• The role of musical and data analysis.
• Data mapping techniques.
• Processes and levels of control the data has.
• Methods to alter or generate sound material and architectural structure.

I will now discuss these elements in detail.

User Interaction and Controllers

As alluded to in Chapter 1 and referenced in the 'Transmutable Music System Diagram' (Figure 2.1), interaction with the system is facilitated through a user interface or controller. Such interfaces range from interface/GUIs, game controllers, musical instruments, and scientific apparatus to more unconventional devices like cars[8] or planes.[9]

While some devices, like game controllers, have pre-set interactions, others, like sensors and a mobile phone's gyroscope or accelerometer, require the developer to define the interactions and determine how the data will influence the music. I will provide more detail about controllers in Chapter 6 on Transmutability.

Data Classification and Transformation

Data, the heart of the control system, can be:

• Imported, received, or autonomously generated.
• Derived from various sources, such as GPS coordinates, APIs, or music information retrieval.
• Textual, numerical, or event-driven.

Furthermore, data can be categorised as discrete and continuous:

Discrete Data

Discrete data refers to distinct, individual pieces of information or events. Unlike continuous data, which flows incessantly, discrete data represents specific moments or occurrences.

Examples & Application in Transmutable Music:

Keyboard Input: Pressing the 'm' key on your computer produces a distinct event that can be mapped to a specific outcome – in this case, the letter 'm' appearing on your screen.

Controllers: Devices like game and MIDI controllers generate discrete data. A button pressed on a game controller, playing a note on a keyboard or a mouse click can be mapped to specific actions or outcomes.

Mapping to Musical Events: Discrete datasets can be linked to various musical occurrences. For example:

• Starting or stopping a loop.
• Playing a specific note or sound.
• Activating a musical parameter or effect.
• Commencing a new section of a track or launching a transitional phase.

Continuous Data

In contrast, continuous data refers to an unbroken stream of information where values change in real-time or periodically.

Examples and Application in Transmutable Music:

• **Sliders and Knobs:** Manipulating a slider or turning a knob on a musical instrument or mixer results in a continuous flow of data. This can correspond to volume, pitch, or other changes in musical parameters.

- **Mouse or Joystick Movement:** The mouse's trajectory or the joystick's tilt can produce continuous data, reflecting the ongoing movement or position of the device.
- **Sensors:** Devices like accelerometers or gyroscopes output continuous data based on their sensing of movement or orientation.
- **API Updates:** Some APIs, like weather services, might provide periodic updates, offering new data every few minutes. Even though the data is received periodically, the continuous nature is reflected in the ongoing updates.
- **Run-time Parameters (RTPs) in Gaming:** RTPs represent continuous data streams, altering game parameters in real time. For instance, the changing distance between a player and a target might be an RTP, which could be linked to the intensity or tempo of background music.

Both discrete and continuous data play vital roles in Transmutable Music. Discrete data often triggers specific events or changes, offering precise control. Continuous data, on the other hand, allows for fluid modulation of music, granting artists the ability to mould and shape soundscapes in response to ever-changing inputs. When designing a Transmutable Music system, understanding the nuances of these data types is crucial to harnessing their potential fully.

Whether discrete or continuous, data can either be input into the system or produced autonomously by the system. Once the data is integrated, it may undergo various forms of transformation. Numerous technical processes exist to transform and utilise this data, as discussed in Chapter 6 on Transmutability. The design of the control system is contingent upon the complexity and type of data and the overarching system design.

There is a vast array of potential data types. Examples include data produced through autonomous processes like algorithmic composition and audio analysis. Additionally, data can be sourced from mobile phone sensors, GUIs, game controllers, VR tracking, GPS sensors, live datasets, contextual data, and more. Many interactive technologies offer standardised datasets, such as mouse or button events, which provide reliable control data. These standardised datasets can be adeptly used to control events and processes in their musical designs. However, outside the realm of gaming, input data often necessitates transformation and classification to generate a more meaningful dataset.

Several processes exist to classify this data, ranging from simple to intricate algorithms. Wooller et al. (2005) delve into the processes of algorithmic music, identifying two primary ways data can be transformed: adjusting the data's potential size and altering its representation scheme (Wooller et al., 2005). The data size might need adjustment to fit the range a system's parameters require. For instance, MIDI data might need scaling to fit a range of 0 to 127. Similarly, for effects like reverb or delay, the required value range could a float[10] between 0 to 1. These are rudimentary examples of data transformation. More advanced transformations could involve stochastic processes to discern patterns in datasets or generate more consistent data insights; see Chapter 6.

Wooller et al. (2005) highlight three algorithmic methods for transforming data for musical mapping: analytic transformation, engine generative algorithms, and learning algorithms. Learning algorithms, in particular, are invaluable for deciphering highly intricate or noisy datasets. Rebecca Fiebrink (2016) offers a comprehensive perspective on how machine learning algorithms can be creative tools in music systems. It is worth exploring her course on machine learning for artists and musicians.[11] She pioneered the software application *Wekinator*, which enables the integration of machine learning into artistic endeavours. Fiebrink discusses how machine learning can achieve five primary musical objectives: recognition, mapping, tracking, unveiling new data representations, and collaboration (Fiebrink et al., 2016).

As technology, especially AI, continues to evolve, our ability to process, interpret, and act upon vast amounts of data grows. However, this commodification of personal data is concerning to me. Individuals generate data footprints with every click, interaction, and online transaction. As this data accumulates, its potential value – both economically and strategically – increases. This scenario is reminiscent of the gold rushes of the past, but with personal data being the sought-after resource.

The challenge is balancing the benefits of these advancements with ethical considerations and privacy protections. Just as the gold rush brought opportunity and exploitation, the 'data rush' presented similar challenges. Ensuring that AI serves as a tool for the betterment of society, rather than just a means of profiteering, is essential as we navigate the next digital frontier.

Audio Analysis

The academic field of music information retrieval (MIR) is a subdiscipline of information retrieval. It is the backbone of many new musical experiences and products. You can think of it as prediction software for creating playlists and recommendations for music, films, games, and TV series. However, its potential is much more significant than mere prediction tools. MIR involves sorting, analysing, retrieving, and categorising music from audio and symbolic data. This area is rapidly expanding with methods to extract and analyse music in real-time. AI, including machine learning, are used, as is Semantic audio. Semantic audio is a way to extract meaning from audio. For example, high and low metaphors represent pitch or the myriad of terms used in mixing, including 'warm up the bass' or the 'vocals are too bright.'

Whether performed in real-time, requiring significant processing, or during the control system development, analysing audio elements and meaningful metadata or descriptors provides novel ways to organise a Transmutable Music system. This is based on identifying distinctive features found within the music. These features can be extracted from audio, like onset values, transients, tempo, style, and feel. Feature-based mixing uses information obtained from an analysis. The control system can then be programmed to make rules and tasks based on these features. For example, only play the tracks with a female singer, do not play sad music, or select loops with limited percussive elements. 'Sad' here is an example of semantic audio where the composer defines the quality of sadness as having a particular combination of notes, durations and tempo. This could go further than this; see the *Semantic Machine* case study for more on feature-based mixing Pg x.

MIR, Semantic Audio, AI and other analysis methods in music offer an extensive range of techniques and software to create advanced systems for Transmutable Music. There are many examples of external objects used for audio analysis in Max 8 and PD, Super Collider plugins, web audio APIs, and programming models/libraries for coding languages like Chuck, python, JavaScript, etc. By combining analysis methods with learning algorithms, a music system could continually compose new music and or music content, which could be customisable to the listener's personal taste.

Control and Process

When developing a Transmutable Music system, the composer must decide:

- What controls the way music can be changed?
- What controls the way the content or music is organised?
- The level of control required by the system or user to change the musical output.

The level of control could be any combination of the previous elements. In my PhD, I categorised Control in a Transmutable Music system by the following.

- User control – the user can, directly and indirectly, control the rendered playback or other elements.
- Machine Control – an autonomous process controls the rendered playback. This includes a system, dataset and/or algorithms developed to control the musical output.
- Chance Control – random processes control the rendered playback.
- Contextual Control – data controls the rendered playback. Data produced by users interacting with the system would be classed as user control. Data control includes contextual data or another system that produces data, for example, data sonification processes (Redhead, 2020).

The level and type of control influence the processes used and vice versa. As discussed in Chapter 6, the processes can be understood as compositional processes or a set of rules and instructions used to compose. Although any process could be used in a Transmutable Music System, I have summarised the main ones.

- Interactive – the work is composed so the user controls some or all the interactions. Audio and/or visuals can be controlled in real time by the direct actions of the user.
- Adaptive – primarily used in gaming and is defined as non-linear music. The music is composed to adapt and takes form to support the gameplay and actions of the user.
- Autonomous – system-led music works independently of direction. This could include generative AI compositions and systems.
- Reactive – context based on and reacting to data or the environment in which the data stream is constantly updated. For example, driving a car controls the playback, like in VW and Underworld's *Play the Road*.[12]
- Responsive/Contextual–adapts music to the user's environment or actions but not through direct actions where the user is in complete control and driving the interaction, as in interactive music. Here, the user's contextual data, for example, location or weather, is used to influence the playback of the music.
- Generative – music is created by a system and is ever-changing.
- Algorithmic – generating music using mathematical approaches, including Markov chains, stochastic algorithms, automata, and highly complex Fourier analysis.
- AI is an area of algorithmic music that uses learning, problem-solving, and knowledge-based algorithms to create music. This includes machine learning and deep learning algorithms.

Mapping

Mapping is a term used to describe the process of using data to control musical parameters. You could think about Transmutable Music as a multiplicity of mappings. So far, in my abstract description of a control system for Transmutable Music, once data has been organised and transformed, it can be used to control compositional processes. In plain terms, this means that the data is mapped to control the musical architecture and or content producing the rendered music output.

We can use well-established practices in computer music fields to guide us in data-based music mapping. In summary, data mapping can be categorised into four types:

- **One-to-one:** a singular output controls a specific input, such as a slider altering an oscillator's pitch.
- **One-to-many:** a single output influences multiple inputs, like one gesture adjusting several synthesis parameters.
- **Many-to-one:** multiple outputs determine a singular input's value (Hunt and Kirk, 2000; Wanderley and Depalle, 2004; Drummond, 2009).

Using interactive technologies involves mapping the data received to control musical parameters. Mapping fundamentally drives the musical design of a

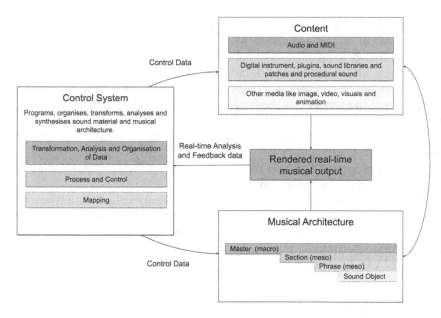

Figure 3.7 Musical Design of Transmutable Music – Control System Added.

Transmutable Music system. In Chapters 6, 10, and 11, I will provide a more in-depth overview of mapping.

Summary of the Control System

Without a control system, Transmutable Music cannot be experienced. It is an essential part of the musical design. To go further with this analogy, the brain of the system and the corresponding nervous system within it control all the parts that make up the whole work. Figure 3.7 provides the completed musical design model described in detail throughout this section.

Summary

The chapter outlines the four essential components of Transmutable Music systems: Content, Musical Architecture, Control System, and Experience.

Multimodal feedback is used to enhance user experience and engagement with Transmutable Music.

A Transmutable Music work should be designed to provide an experience beyond a static form. The design should produce some level of variability in its playback and feedback so that changes in the music can be recognised.

Considerations for designing a Transmutable Music experience include listener engagement, narrative interactivity and control, creative direction and composition, and technical and system considerations.

The musical design for Transmutable Music is outlined, focusing on the interaction between the control system, content, and musical architecture.

The content in Transmutable Music describes a variety of materials, including sound, images, videos, animations, sound libraries, patches, digital instruments, MIDI, procedural sound and code.

The musical architecture in Transmutable Music defines how the control system alters and transmutes the content and musical architecture on various hierarchical levels.

The hierarchical levels include the master, section, phrase, and sound object levels.

The control system manages the transformation, mapping, and organisation of music and data within the system.

When building a control system, important elements include data type and transformation, musical and data analysis, data mapping techniques, control processes and levels, and methods for altering or generating sound and architectural structures.

Control can be categorised into four types: user, machine, chance, and contextual control.

Mapping is a term used to describe the process of using data to control musical parameters.

Think of Transmutable Music as a multiplicity of mappings.

Exercise 3.1 Crafting Your Blueprint for a Transmutable Music Work

As we conclude our exploration of Transmutable Music systems, let us apply our understanding in a practical context. This exercise is designed to guide you through the initial planning stages of a Transmutable Music project.

Objective: Develop a foundational plan or blueprint for your Transmutable Music Work, focusing on the artistic goals, processes, control mechanisms, and the listener's experience.

Instructions:

1 **Conceptualisation:**

   - **Artistic Goal:** Reflect on the primary artistic objective of your work. What are you aiming to express, discover, or innovate?
   - **Experience Focus:** How do you envision the audience interacting with your work? Is the audience's role passive, active, or somewhere in between?

2 **Control Mechanisms:**

   - **Types of Control:** Will your work employ user control, machine control, elements of chance, or contextual data? Refer to Chapters 2 and 3 for a detailed understanding of each type.

3 **Process Selection:**

- From the list that follows, select the processes that will drive your work. You may choose one or combine several to achieve your artistic vision:

  - Interactive
  - Adaptive
  - Autonomous
  - Reactive
  - Contextual/Responsive
  - Generative
  - Algorithmic
  - AI

4 **System Design:**

- **Self-Designed System:** If you are creating the system, sketch a schematic that illustrates its functionality. Consider the content, the hierarchical structure of the musical architecture, and how the user will interact with the system.
- **Pre-Designed System:** If utilising an existing system, outline how it functions and how it will serve your project.

5 **Experience Extension:**

- How will the transmutable work expand the listener's experience beyond what static music offers? Consider the novelty, depth, and engagement your work will provide.

Reflection:

- Upon completing these steps, you should have a clearer vision of your Transmutable Music project. Reflect on the relationship between your artistic goals, chosen processes, and the proposed user experience.

Output:

- Compile your responses into a project blueprint. This document should articulate the core concept, chosen processes, control types, system design, and anticipated user experience.

Sharing:

- Consider sharing your blueprint with peers or mentors for feedback. Other people's perspectives can offer valuable insights and help refine your concept further.

Next Steps:

• With your blueprint in hand, you are now equipped to move forward into the prototyping phase, where your concepts will begin to take shape in the real world.

Notes

1 https://www.merriam-webster.com/dictionary/system
2 https://www.ninjajamm.com/
3 MAO is 'a Semantic Web framework that investigates new ways in which music can be experienced on mobile devices.' (Thalmann et al., 2016b)
4 Dymos are playable semantic web graphs. They are multi-hierarchical structures and constraints' (Thalmann et al., 2016a).
5 https://www.ninjajamm.com/
6 https://www.nagualsounds.de/
7 https://www.facebook.com/weavmusic/
8 Underworld, Play the Road https://www.youtube.com/watch?v=3flwZ8OpXBY
9 Bob Jarvis, Aileron One https://www.youtube.com/watch?v=Ct-Sb-zGlG0
10 A float or a floating-point number is a term used in computer programming to describe a number that has a decimal point. Floating-point numbers are used when more precision is needed than integers (1,2,3, etc) can provide.
11 http://www.wekinator.org/kadenze/
12 Underworld, *Play the Road* https://www.youtube.com/watch?v=3flwZ8OpXBY

References

Drummond, J. 2009. "Understanding Interactive Systems." *Organised Sound* 14 (2): 124–133. http://search.proquest.com.ezproxy.newcastle.edu.au/docview/1554438?accountid=10499.

Fiebrink, Rebecca, Baptiste Caramiaux, R Dean, and A McLean. 2016. *The Machine Learning Algorithm as Creative Musical Tool*. Oxford, United Kingdom: Oxford University Press.

Hodl, Oliver, Geraldine, Fitzpatrick, and Simon, Holland. (2014). "Experimence: Considerations for Composing a Rock Song for Interactive Audience Participation." In *Proceedings of ICMC/SMC Joint Conference: 40th International Computer Music Conference and 11th Sound and Music Computing (ICMC/SMC 2014)*.

Hunt, Andy, and Ross Kirk. 2000. "Mapping Strategies for Musical Performance." *Trends in Gestural Control of Music* 21 (2000): 231–258.

Lee, Sang Won, and Freeman Jason (2013). "Echobo: A Mobile Music Instrument Designed for Audience to Play." In *the Proceedings of New Instruments for Musical Expression (NIME)*, Ann Arbor, MI 48109-2121 1001.

Redhead, Tracy. 2020. *Dynamic Music: The Implications of Interactive Technologies on Popular Music Making*. NSW, Australia: University of Newcastle. http://hdl.handle.net/1959.13/1413444.

Roads, Curtis. 2001. *Microsound*. Cambridge, USA: MIT Press. http://ebookcentral.proquest.com/lib/newcastle/detail.action?docID=3339387.

Roads, Curtis. 2015. *Composing Electronic Music: A New Aesthetic*. USA: Oxford University Press.

Scipio, Agostino Di. 1994. "Formal Processes of Timbre Composition Challenging the Dualistic Paradigm of Computer Music." In *Proceedings of the International Computer Music Conference*, 202. International Computer Music Association.

Spiegel, Laurie. 2000. "Music as Mirror of Mind." *Organised Sound* 4 (3): 151–152. 10.1017/S1355771800003046.

Thalmann, F, A P Carrillo, G Fazekas, G A Wiggins, and M Sandler. 2016a. "The Mobile Audio Ontology: Experiencing Dynamic Music Objects on Mobile Devices." In *Proceeding of 2016 IEEE Tenth International Conference on Semantic Computing (ICSC)*, 47–54. 10.1109/ICSC.2016.61.

Thalmann, Florian, György Fazekas, Geraint A Wiggins, and Mark B Sandler. 2016b. "Creating, Visualizing, and Analyzing Dynamic Music Objects in the Browser with the Dymo Designer." In *Proceedings of the Audio Mostly 2016*, 39–46.

Wanderley, Marcelo M, and Philippe Depalle. 2004. "Gestural Control of Sound Synthesis." *Proceedings of the IEEE* 92 (4): 632–644.

White, Greg. 2015. "Towards Maximal Convergence: The Relationship Between Composition, Performance, and Production in Realtime Software Environments." NSW, Australia: University of Newcastle. http://hdl.handle.net/1959.13/1312716.

Wooller, R, Andrew R Brown, Eduardo Miranda, Rodney Berry, and Diederich Joachim. 2005. "A Framework for Comparison of Processes in Algorithmic Music Systems. Generative Arts Practice." *Creativity and Cognition Studios Press* Generative: 109–124. http://eprints.qut.edu.au/6544/1/6544.pdf.

Section 2

Approaches for Composing Transmutable Music

4 Experience and Interaction

Overview of Section 2

As I consistently emphasise throughout this book, composing music that can change at any time completely differs from composing music that is static or fixed. The process is far more complex; the structure and form of the work operate within different mediums. To better understand the concept of Transmutable Music, let us consider the analogy of an artist painting a picture on a canvas and then scanning it into a computer to digitise it. The expectation that the painting will become animated after digitisation is wrong. The scanned image will only be a static form, such as a PDF file. Painting a picture and creating an animation are two different processes involving two different mediums.

Similarly, creating Transmutable Music requires different skills than those needed for traditional songwriting or composition. Mastery in one does not necessarily translate to proficiency in the other. That said, many foundational skills and concepts in music composition are transferable and can serve as a base for both mediums. In the following four chapters, I will discuss practical processes and strategies for composing music for the unique and innovative medium of Transmutable Music.

Three foundational approaches will be discussed: Transmutability, Variability, and Experience.

- *Transmutability* specifies the manner and extent to which control data is employed to alter the musical form of a piece.
- *Variability* pertains to the method and degree to which variation is intentionally composed into the music, ensuring that each iteration of the piece offers a unique auditory experience.
- *Experience* is used in the design and testing of the work. Ensuring it works from a conceptual and functional perspective.

These offer systematic approaches to both the composition and evaluation of Transmutable Music. This enables Transmutable Music to realise its potential and provide fresh, novel experiences for audiences.

DOI: 10.4324/9781003273554-6

Given the fluid and dynamic nature of Transmutable Music, composers must craft their pieces with these approaches at the forefront of their creative process. Furthermore, these concepts serve as invaluable tools for evaluating Transmutable Music, facilitating the identification of avenues to enhance and enrich the listener's experience.

The Paradox of Experience in Transmutable Music Systems

Transmutable Music is directly related to experience. When listening to the rendered musical output of Transmutable Music, a moment in time is perceived. This moment is an instance of all the possible combinations and processes constructed within the system. This moment or rendered musical output is the outcome of the overall form of the Transmutable Music system, which challenges the traditional concept of fixed forms.

Although the rendered musical output or one instance can be analysed within these traditional concepts, every time the system creates a new output, it will be different. Since listeners are unaware of all the different ways the music may be structured, they are only listening to one instance. Each instance has a duration and is perceived by the listener as a fixed form (static) for that moment. As a result, the listener can only hear the resulting musical output they are experiencing or interacting with.

This raises a crucial question: if the music is perceived as unchanging, what is the purpose of creating a Transmutable Music system? The crux of this matter lies in the experience. Users need to grasp that the music is capable of transformation. How a user interacts with and perceives the music hinges on the degree of control they are afforded within the system. The experience of Transmutable Music can be enriched through several methods, such as incorporating feedback mechanisms that signal to the listener that the music can evolve, or by ensuring the music varies with each play, emphasising its dynamic nature.

Some Transmutable Music works have the listener active and in the centre of the compositional process, like Ninja Jamm.[1] Others demand no interaction from the listener, like Brian Eno's generative app Reflections.[2] In Transmutable Music that does not require audience interaction, '... it is the variability of playback in the music that provides a new experience for the listener. The music is always unique and infolding, it does not represent any other moment in time than the now' (Redhead, 2020, 106).

For Transmutable Music works where the audience plays an active role, it may be important to understand how audiences might perceive, experience and understand the work. This understanding can be achieved through user studies. We will be discussing *Variability* in the next chapter; however, for this chapter concerning *Experience*, what are the important considerations needed for designing a Transmutable Music Project, given it must offer an extended musical experience?

Integrating Continuous Evaluation in Transmutable Music Composition

Evaluating experience within a Transmutable Music system/work is not something that is performed once you have finalised your project. Think of it more as a tool that can be used in the compositional and design process. As a composer or producer, you constantly evaluate your work. You objectively listen and ask yourself: Do the sections transition smoothly? Is this a good vocal take? Does it have the right emotional intensity? Is it an authentic performance? Does my track need more bass (yes, always!)? You are constantly making decisions about how your work is evolving. Evaluating experience can be thought of as an extension of these processes and personal approaches. However, as an artist, you learn to trust the aesthetic choices and processes you use to compose a work, but you cannot possibly know how other people will respond and understand the work. There are Tools used in music to engage and enhance emotional responses in listeners of static music. For example

- approaches in lyric writing to help engage the listener's experience, such as the use of imagery to evoke sensorial impressions and mental images, and
- tools to produce suspense and tension from listeners using ominous ambience, increasing tempo, ostinato strings, dissonance, etc.

You could think of evaluating user experience as a similar tool to help your work extend your listener's experience. Just as you develop your own methods to make compositional, performance and production decisions, when working in Transmutable Music, you need to learn how to evaluate and make decisions about the user experience. You cannot instinctively achieve this; you need to go beyond your perception and evaluate from different people's perspectives.

Evaluating user experience has foundations in the Human-Computer Interaction (HCI) and Music Interaction fields and is essential in all design stages. In order to avoid frustration and confusion, the experience must be designed with the aim of engaging and immersing the listener. This can be performed after a significant milestone, like a proof of concept or a prototype. If the project is user-focused and has a level of user control, I recommend some user testing.

It can feel unsettling for musicians to research their audience; however, some testing may be required if the audience roles involve producer-led transmutations. After working in this area for some time, it has become apparent to me that users and audiences do not always (in fact, ever) behave in the way one thinks they will. For example, if your work consists of a red button in the middle of a room, you can guarantee there will be people who will keep pushing the button repeatedly as quickly as possible. Not because they necessarily disrespect your work but out of the desire to see what can be experienced, what the boundaries of the work are, or if they can break it.

To delve deeper into this concept, I will recount a brief case study of a project that explored technology-mediated performance. This initiative was part of the *Nightline Program* at the Ars Electronica Festival in Austria. Since the festival has been a platform for media art since 1979, it seemed an apt setting to investigate producer-led transmutations within live performances. My expectations of audience engagement were upended, to say the least. The outcome was nothing short of chaotic, aptly reflecting the performance's title, *The Madness of Crowds*.

Case Study: *The Madness of Crowds* – Audience Interaction with Technology in a Performance Setting

In 2016, an intriguing blend of creativity and technology came alive at the Ars Electronica Festival through an experimental performance titled *The Madness of Crowds*. This performance was an exploration and a confluence of distinct technologies and innovative ideas brought together to gauge audience engagement in a novel way.

The journey towards this experiment began a year earlier, in 2015, at the Music Tech Fest in Northern Sweden, where serendipity led me to Hakan Lidbo, a prolific interactive music producer from Stockholm. Having been a part of numerous ground-breaking projects, Lidbo's reputation is impressive. Among his ventures, the *World's Largest MIDI Controller* concept stood out as a captivating innovation. It was a dynamic tool that allowed audiences to choose loops for songs. I also met the audio-visual and interactive artist Synthestruct through the Music Tech Fest community and was impressed by her work with cymatics and audiovisuals. Finally, Sydney-based artist Trevor Brown and I have been looking at ways to collaborate. He is an incredible musician who explores music technology and performance.

Thus, with a spark of an idea and a gathering of like-minded artists, *The Madness of Crowds* was conceived to explore how one might transform the conventional passive concert experience into an immersive interactive spectacle.

The Madness of Crowds was performed on 9/9/2016 at Salon Stage POSTCITY at the Ars Electronica Festival in 2016. The performance was experimental by nature and a collaboration between Hakan Lidbo, Synthestruct (Ginger Leigh), Trevor Brown, and myself.

Objective

This performance aimed to experiment with a Transmutable Music system in a real-time performance environment. We wanted the audience to play with the music and be part of the performance. Trevor Brown came up with

Figure 4.1 Picture Taken at Ableton Loop 2015 of the *World's Largest MIDI Controller.*

the title *The Madness of Crowds,* which was a perfect fit to describe the performance. We will never know if the title gave permission to the audience to go mad or if it was the manic music produced that caused a chain reaction.

Project Details

Central to the performance were the *World's Largest MIDI Controllers* (see Figure 4.1) – three giant cubes. These were interactive instruments, allowing the audience to influence the music directly. Each cube, designed by Hakan Libdo and Per Olov Jernberg, had unique coloured symbols on each side, representing drums, bass, or melody. As the audience adjusted the cube sides, corresponding loops were triggered. This interaction was meant to transform the audience from mere spectators to performers. I saw the *World's Largest MIDI Controllers* at the Music Tech Fest (Umea, Sweden) and Ableton Loop (Berlin) 2015. I witnessed the audience thoughtfully using the cubes to compose at each event, working together, and enjoying the experience. Combined with the *World's Largest MIDI Controllers*, the audience could also drive the performance's pulsing interactive and audio-reactive projections using – a dance visualiser designed by Synthestruct called *react( ).*

Implementation

The performing space had two large screens erected to enhance the audience's experience. The first showcased *react( )*, and the second was a projection of the patterns from the three cubes. This visual cue was intended to help the audience match the patterns to the sounds being produced. The ultimate goal was for the audience to learn and play the cubes as instruments.

The technical setup was intricate. Sensors embedded in the cubes sent data to an Ableton live set through Bluetooth. This setup, orchestrated by Lidbo and Jernberg using sensor tags designed by Texas Instruments, allowed for real-time music adjustments based on cube orientation. A detailed hierarchy of the music's design was established, with each song containing 18 instrumental loops divided among the three cubes. I provide an evaluation of the transmutability and variability of this project in Chapter 5. Trevor Brown and I edited and mixed the loops in the studio the days before the event.

Performance

The 30-minute spectacle featured five original compositions, with the audience in control of selecting musical loops. Trevor Brown (on saxophone and live sound processing) and I (providing vocals and effects) performed live, reacting to the audience's selections.

Outcome and Learnings

Suffice to say, the event did not go as expected. Instead of the anticipated thoughtful interactions with the cubes, the audience, perhaps due to the late hour and the presence of alcohol, threw the cubes around the space, resulting in rapid and erratic changes in the music. This chaos was far from the intended audience experience.

Despite this unforeseen interaction, the performance was adaptive. The live improvisation by the performers provided a semblance of control and structure to the otherwise chaotic musical output. Given the experimental nature of the festival, the music suited the event well, and the audience was having a great time using the MIDI controllers like a game of volleyball.

Key Observations

Audience Culture Impact: The unpredicted audience behaviour highlighted the evolving nature of audience culture in technology-mediated performances. Unlike traditional performances with established audience etiquettes, interactive media performances are still carving out their norms. Take, for example, a classical music concert; the audience will sit quietly and listen. In a club setting, people will dance and talk. At a popular live music concert, the audience will dance, talk, scream and sing along. All these examples have a known or even a priori knowledge of the correct social edict. For example, audience members do not dance and scream at a classical orchestral concert.

In the same way, people sitting silently and listening intently in a club setting may create looks of equal disgust from other audience members. Within society, we all learn the correct social constructs; however, interactive or technology-mediated performances do not have the same social knowledge. They are new and evolving, and audiences do not know how to behave. Interestingly, this might have contributed to why this performance was a fun yet slightly manic and out-of-control event.

Event Environment: The festival's atmosphere, the time of the event, and the fact that it has been a hub for new media since 1979 all influenced the audience's reaction. Even with instructions, their approach to the interactive elements varied significantly.

Future Recommendations: For upcoming performances, understanding the audience's familiarity with interactive media and adjusting the environment or tools accordingly could yield different outcomes. Multiple performances in varied contexts might offer a broader understanding of audience reactions. I would also compose and produce a variety of different interaction options in order to adjust the performance's representation of the different ways the audience might interact at each event.

Conclusion

The Madness of Crowds was an ambitious endeavour to blend technology with art in a live setting. Though the audience's interaction diverged from the initial intent, the performance provided invaluable insights into the intersection of technology, music, and audience engagement. It emphasised the need for adaptability and a deep understanding of audience culture in interactive performances.

The Importance of Observational Insights

Our case study on *The Madness of Crowds* performance serves as a compelling narrative on the unpredictability of user interaction. Despite the absence of formal user testing, keen observations of the festival-goers revealed a wealth of insights. It is a classic example of the unexpected ways people might engage with a project. Observing a crowd's natural interaction with an installation can often be as illuminating as any structured experiment. Such was the case when unprompted attendees interacted with the performance in ways that defied our expectations, reminding us that the most valuable feedback often comes from simply watching and learning.

In the realm of music interaction and human-computer interaction, past research can provide existing frameworks for many types of music works. For instance, the evolution of mobile phone use offers a fascinating story

(Metaswipes, 2023). The now-commonplace swiping gesture was born from extensive user testing to understand the most natural movements for browsing content. This anecdote highlights the significance of rigorous user testing: a feature that feels 'instinctive' to millions worldwide was meticulously crafted through trial and error.

Another illustrative example is how toddlers can navigate tablets without prior instruction, demonstrating the triumph of intuitive design. It is a testament to the extensive research and testing that has honed the user experience to align with our innate cognitive abilities. However, these successes do not eclipse the individual differences in perception and learning; what is intuitive for one may not be for another. The quest in human-computer interaction research is to find a balance that caters to the broadest user base while acknowledging these variances.

Prototyping and Iterative Development

When it comes to user testing, the journey from concept to execution is a delicate dance of hypothesis and experiment. Prototyping is an essential model for developing video games and software, which involves testing at different stages of development. This testing may only sometimes include user testing. At earlier stages of development, a project may not be ready for user testing. Developers or even friends and family can evaluate it until it is at a stage where user testing is required and will be most beneficial. By using the criteria provided for Transmutable Music systems in Chapter 7, composers and producers can evaluate their systems effectively using Variability and Transmutability. This can assist in developing a solid prototype on which to continue development. In game development, there are often multiple stages of user studies to refine the user experience over years of development. Prototyping enables developers to test the waters with their ideas and refine them with each iteration.

In the creative process of music interaction design, the initial trust in one's vision must be balanced with flexibility for change. Composing for different mediums is very difficult if the work is not experienced within that medium. For example, a game composer may find that a score that seemed perfect in isolation does not quite fit the dynamic environment of the game. These scenarios underscore the importance of revisiting and revising one's work, guided by user experience.

Advanced Prototyping using iterative development is essential for composers and producers of Transmutable Music Works. It is important to note that along with the artist's conceptual vision for the project, the designing of variability and transmutability are equally crucial in realising the final work.

User Testing

When integrating user testing into the development of Transmutable Music projects, it is essential to consider how it can be embedded effectively to enhance the creative process and end-user experience. This section will outline methods to incorporate user testing at various developmental stages that are tailored to the unique context of Transmutable Music.

Initiating user studies requires a critical assessment of their relevance to your project. You should consider the established UX best practices, the significance of user interaction within your project, and whether the insights gained from the study can be feasibly implemented within your resource constraints.

User studies range from small-scale pilot studies, manageable and insightful for independent artists and academics, to extensive testing for projects with larger commercial ambitions or significant research funding.

In Transmutable Music, user testing can include a blend of qualitative (descriptive) and quantitative (statistical) methods, commonly known as mixed methods. This discussion will not provide a deep dive into research methodologies but will highlight suitable methods for Transmutable Music projects, such as surveys, focus groups, observations, and interviews. The critical issue is identifying or understanding the problem or issues you want to address in your user study. This can also be called the research problem.

A well-designed research plan is crucial for meaningful analysis (Creswell, 2009). Start by identifying the issues your project aims to address. For example, *The Madness of Crowds* case study explored the integration of interactive technologies into live performance, revealing that audience interactions were unpredictable and could deviate from the composer's vision. Such observations are vital for subsequent system redesigns to improve user control over the musical output.

When defining the research problem, articulate a set of focused questions to direct your study towards precise outcomes. For Transmutable Music, consider these critical questions:

1 **User Understanding and Influence:** Do users comprehend their role in influencing the musical output? Interactive systems need to provide clarity and empower users.
2 **Enhancement of Listening Experience:** Does the system add value to the conventional listening experience by encouraging deeper engagement?
3 **Accessibility and Ease of Use:** Is the system designed to be user-friendly and accessible to a diverse range of users, regardless of their musical or technical background?
4 **Promotion of Engagement:** Does the system invite users to explore and utilise its full capabilities?
5 **Aesthetic and Emotional Connection:** How do users emotionally connect with the thematic elements of the work?

Before any research, it is prudent to investigate if similar studies have been conducted to leverage existing insights and avoid redundancy.

According to Dawson (2006), begin with the five Ws – what, why, who, where, and when (Dawson, 2006). These will guide your research framework.

'What' is your research? Try to sum this up in one sentence. Dawson explains that your research is too broad if you cannot sum this up in one sentence. The 'Why' is to understand usability or another aspect of user interaction. 'Who' are your participants? 'Where' will you conduct the research? 'When' will it be feasible for both you and the participants? (Dawson, 2006)

Ensure your problem statement or research question is concise and manageable. For example, rather than aiming to engage every potential listener worldwide, narrow your focus to something like 'How can I make my Transmutable Music project more engaging for my target audience?' or specifically 'How can I make my Transmutable Music project more engaging for fans of my record label?'

Identifying your target audience is a fundamental step before selecting a research method. The characteristics of your audience will influence which method is most appropriate for your study.

Identifying your target audience is challenging, especially when your release goes worldwide, entering a saturated market. This global release is a double-edged sword – a great opportunity that demands differentiation (Chen et al., 2021). Musicians face this each time they drop a single on platforms like Spotify or YouTube. *The Madness of Crowds* project, for instance, targeted participants at the Ars Electronica Festival, hypothesising that the festival's cultural context might predispose attendees to a better understanding of technology-mediated performance.

Practicality in participant recruitment is critical. Offering incentives like gift vouchers or prize entries can help attract participants, but securing volunteers is challenging, even with the most intriguing projects. How many surveys, for instance, have you participated in? Reflect on your own engagement with surveys to appreciate the challenge.

Once your target audience is pinpointed, the next step is reaching them effectively. Social media networks may suffice for some, but this can introduce sampling bias.

A rigorous research design is imperative for studies aiming for publication or formal presentation, including ethics approval, addressing sampling bias, and safeguarding participant welfare.

Sampling is a critical component of research design, involving probability sampling for generalisable findings and non-probability sampling for more targeted inquiries (Babbie, 1990).

Stratifying your sample group (i.e., identifying sub-groups) helps ensure they match your demographic criteria. Defining the size of your sample group is also critical – this could range from women aged 18-24 with a music background to a broader demographic if your project has a wider appeal.

Your research design should pivot around key questions that define what you need to know and the best methods to obtain that information. Consider the evaluation questions mentioned earlier, which are tailored to Transmutable Music to guide your approach:

- *User Understanding and Influence*: Does the user understand their impact on the music?
- *Extension of the Listening Experience*: Does the system deepen the listening experience?
- *Accessibility and Usability:* Is the system widely accessible and user-friendly?
- *Engagement and Exploration*: Are users motivated to fully engage with the system?
- *Aesthetic and Emotional Response:* How do users emotionally connect with the work?

Designing user tests is intricate and should be custom-tailored. The nature of your project – be it usability, the type of experience, perception, emotional response, or understanding – will dictate the selection of test participants and methods. If your app is for music producers, include industry professionals in your test group. For a general audience, ensure a broad spectrum of musical and technical backgrounds.

Securing participation in user studies can be challenging, and it is crucial to be mindful of sampling bias. Your efforts in the design, recruitment, and implementation of user tests can significantly influence the success and relevance of your Transmutable Music project.

Choose your research method wisely. Interviews offer deep personal insights; focus groups capture collective opinions; surveys can provide specific and broad feedback; and participant observation can yield a nuanced understanding of user behaviour in natural settings.

In essence, user testing is not merely a final checkpoint but a series of explorative engagements throughout the project's lifecycle. Through observation or structured experimentation, the ultimate aim is to create an intuitive and impactful user experience. I will provide two short case studies of pilot studies I have conducted on Transmutable Music projects.

Case Study: 'The One Drop' Project – Interactive Listening Experiences – Will Audiences Interact (2010–2014)

Introduction

As part of my Master of Art's in research, I wanted to explore the idea of an amorphous song. The research culminated in the track – *One Drop*. The creative journey is discussed in Chapter 5. *One Drop* was an experimental prototype composed to explore new forms of engagement with music, allowing listeners to interactively shape and personalise their music experience. It is an example of Transmutable Music as it is interactive and lets the audience remix their own structured version of the song.

Project Summary

The *One Drop* project was rooted in fostering user engagement, drawing inspiration from gaming, where players are not just passive consumers but active participants. The objective was to transform the listening experience from static to dynamic interactive engagement. The work consisted of 18 stems composed to enable users to remix their own version of the track. There was not one version of the song. The song worked with all 18 stems playing and in any combination. For example, the song could work with just guitar and vocals, for instance or, disco beat, electro synth and synth bass. For more details on the composition of the stems and my approach to variability, see Chapter 5.

Research Problem

When researching interactive music technologies produced in 2010, they were limited to the stems of fixed song versions. I found the options to have more control over the music very limiting because there were few creative options to explore. I wanted to compose a work with this format in mind. I composed a song with many ways of being played back, allowing listeners to personalise their experience. I wanted to see whether music enthusiasts might be interested in engaging with a listening experience like this.

Research Design

The project's first phase focused on creating a set of musical works/experiments and an interactive music release format. The results of this part of the study can be accessed in the Art of Record Production Paper[3] (Redhead, 2015). However, we are interested in the user study, which was the project's second phase.

The second phase sought to understand user engagement through qualitative and quantitative methods. The primary goal was to answer the following user-centric research questions:

1 Are audiences willing to engage with music applications that embody principles from remix culture?
2 What factors influence the depth of user engagement and interaction?
3 What observable user interactions can guide future design improvements?

Given the lack of direct literature on audiences' propensity to engage with an interactive music format at the time, this phase was crucial for providing empirical validation.

Based on Dawson's practical guide, I have attempted to retro-engineer the five Ws (Dawson, 2006).

• What – Conduct a preliminary study to observe and survey music enthusiasts, assessing their engagement and interaction with the *One Drop* music prototype. To understand the factors influencing their user experience.

- Why – This study addresses the significant upheaval in the recorded music industry caused by the internet, which has profoundly impacted artists, production companies, and listeners worldwide. The aim is to explore whether interactive listening experiences might uncover new value for musicians.
- Who – Technologically savvy music fans.
- Where – The University of Newcastle, Australia, where participants would use the app and be observed.
- When – Over one month during the semester to attract student participation.

Research Methods

Participants were given 20-30 minutes to explore the app on an iPad and record a version of the song *One Drop* (The *One Drop* Max patch can be found on the Companion Website). They were observed during this process, and their interaction was recorded in the Ableton Live set. Data collection methods encompassed participant observation, an online survey, and audio recordings of the user-generated remixes.

The survey was designed to capture various factors that could influence participants' interactions with a new music interface and their perceptions of its viability in the music consumption market. According to Longhurst, 'a range of different aspects of everyday life ... intertwine in the production of a wide range of interacting audience behaviours.' (Longhurst, 2011, 78) Therefore, variables such as demographic information, income, music listening habits, fandom, musical preferences, technological proficiency, and online activity were included in the survey design.

To gain insights into audience engagement and music consumption preferences, it was necessary to ask some sensitive questions, particularly about the primary cause of disruption in the recorded music industry. Illegal file-sharing was pinpointed as a significant contributor to the industry's decline (IFPI, 2012) at this time, making it a central topic in understanding consumption practices. This line of questioning allowed me to explore correlations between the potential market value and the participants' legal and illegal music consumption habits.

Income data was vital for determining participants' socioeconomic status and for performing correlations between their willingness and ability to purchase recorded music in the format being tested. Such correlations were crucial for assessing whether the format had commercial potential and was worthy of further research and development. Questions about income are standard when examining music consumption practices. For example, Woodward (2011) notes that 'an individual's level of spending on going to live music concerts might be explained by a range of factors such as discretionary income, the time available for leisure, and proximity to live music venues' (Woodward, 2011, 958). Statistical analysis can then reveal how strongly these factors influence concert attendance, allowing researchers to understand consumer patterns.

Participants

The pilot study enlisted 40 participants, ensuring gender balance and targeting demographic groups known for substantial digital music consumption. Criteria included age range (dominated by 18-49-year-olds) and recent attendance to over five live music events, signalling active music fandom.

A multifaceted recruitment strategy was employed to source participants, involving a Facebook ad campaign, physical posters at music venues, and outreach to the University of Newcastle Conservatorium's student body. Participants were given a $20 voucher for participating.

Instruments

Due to the experimental nature of this work, I wanted to prototype the composition and gain users' feedback on whether this type of listening experience could gain popularity.

TouchOSC was used to design the iPad's GUI (Graphic User Interface). This interface allowed users to manipulate and experiment with the musical stems in real time, sending feedback to Ableton Live.

Two versions of the interface were created – simple (Figures 4.2) and advanced (Figures 4.3). These two interface options gave users varying degrees of control, from basic toggling of stems to intricate volume and effect manipulations (Figure 4.3).

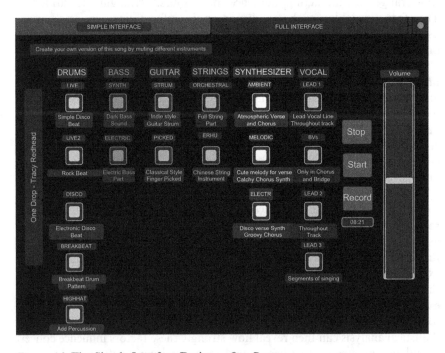

Figure 4.2 The Simple Interface Design – *One Drop*.

Figure 4.3 The Advanced Interface Design – *One Drop*.

Data collection tools were diverse, including self-reporting, online surveys (Survey Monkey), observational notes, and audio recordings. Key Survey software streamlined data entry and collection, while a combination of TouchOSC and Ableton Live captured participant interactions in detail during their remixing sessions.

Analysis

I used a combination of storytelling and number crunching to understand the data better. I organised the information to make it clean and ready for analysis. Then, I took the detailed feedback and experiences people shared, which are usually hard to measure, and converted them into categories and themes that could be counted and compared. This helped me see patterns and compare what people said in surveys and what happened in the audio recordings. For instance, I looked closely at how people changed the music tracks and used different sound effects and compared that with their musical and technical knowledge. I used Microsoft Excel to analyse the data and make comparisons.

Results

Here is a summary of the user study results in relation to the research questions asked.

1 Are audiences willing to engage with music applications that embody principles from remix culture?

- **Positive Reception:** The study shows that most participants found value in the app, with a significant majority (95%) liking its use and a high average rating (87.5%), indicating that audiences are willing to engage with such music applications.
- **Engagement Without Prior Experience:** Despite many participants (62.5%) having never DJ'd or remixed music, they engaged with the app, demonstrating its appeal to a broader audience.
- **Monetisation Potential:** Despite a contradiction with some admitting to illegal downloads, the willingness to pay for the app or its downloads suggests that users see enough value in these applications to monetise them.

2 What factors influence the depth of user engagement and interaction?

1 **Technology Proficiency:** The fact that most participants owned personal computers and spent considerable time online indicates that tech-savvy users are more likely to engage deeply with web-based music platforms.
2 **Advanced Control Preference:** A significant portion (65%) used the advanced interface to record their version, hinting that users prefer advanced controls to engage more deeply.
3 **Music Interaction Background:** Users with a background in music-based games showed a higher appreciation for music apps, suggesting that past interactions with music influence engagement depth.
4 **Professional Feedback:** Some professionals found the app limiting, indicating that more experienced users may seek depth regarding features and creative freedom.

3 What observable user interactions can guide future design improvements?

- **User Interface and Interaction:** Participants desired more visual and interactive features, such as a track bar, visual waveforms, and the ability to add sounds or loops. This feedback can directly inform UI/UX design improvements.
- **Guidance for Users:** The study revealed that some users felt overwhelmed and needed additional guidance, suggesting the inclusion of tutorials or help features in the future design.
- **Diverse Engagement Levels:** The range of engagement, from those making no adjustments to those experimenting with multiple features, indicates the need for a design that caters to both novices and advanced users.
- **Artistic Variety:** The creator's reflection on the limited diversity in arrangements and vocal choices underlines the need for features that encourage more creative exploration within the app.

Figure 4.4 Semantic Machine Web Application Browser Image.

Case Study: The Semantic Machine – Redefining Music in the Age of Interactive Technology

Introduction

The *Semantic Machine* (see Figure 4.4) is a mobile application that leverages semantic web technologies to craft a listening experience that's inherently personal and contextual. Instead of a fixed song, listeners encounter a musical narrative that evolves based on their surroundings, from location and time of day to prevailing weather conditions.

Project Summary

The *Semantic Machine* a fictional concept based on the datafication of human experience. The *Semantic Machine* is a metaphor for surveillance capitalism. This system comprises your personal data that you give away for free whenever you use your computer, smart devices, or wearables. The work is an ever-changing song that is influenced and controlled by the user's location, weather, and time of day. Conceptually, the *Semantic Machine*, a fictional all-knowing AI machine, is central to this artwork. The lyrics of the work present an argument of control between a human and AI, from contrasting perspectives. It warns of surveillance-driven business models undermining our autonomy (Zuboff, 2019).

The *Semantic Machine* speaks to us in a dark tone that it knows us better than we know ourselves. It warns us that we are on the edge of hindsight and that monetising our personal data, experience, and inner thoughts is a plague in our society. The machine tells us to ignore what is happening and carry on, to dance away all our fears of doomsday. The song changes, just like we all do, based on the weather, time of day and location. It is as if the work has a mind of its own.

Developed in collaboration with Florian Thalman as part of the FAST research project,[4] the Semantic Player underpinning the app is a testament to the prowess of semantic technologies. These technologies, rooted in AI, discern meaning and relationships within intricate data structures. As listeners access the app in real-time, it harnesses their GPS data, interacts with external APIs, such as Weather API and Google Maps, and then musically interprets this data. From temperature to cloud cover, each environmental factor informs a distinct musical response, ensuring that every listening session remains unparalleled.

However, despite its innovation, the *Semantic Machine* does not claim to supplant traditional music forms. It simply introduces a novel dimension to musical exploration. From a composer's perspective, the *Semantic Machine* is an avenue to engage with multiple musical ideas and perspectives simultaneously.

Research Problem

The primary objective of this project was to gauge the viability and reception of context-driven music consumption, i.e., music that changes based on certain variables like time, location, and weather.

This project addresses large-scale, unprecedented disruption in the recorded music industry caused by the internet and digitisation and the direct challenges it has presented to traditional music production and distribution modes. The significant economic implications have repercussions for artists, companies, and audiences. Hence, a survey is used to get an overview of the necessary variables that might influence participants' interactions with the work and the perceived viability of contextual listening experiences in a music consumption marketplace.

Research Design

Two primary phases characterised the project's research approach:

Phase 1 – This phase culminated in the creation of the *Semantic Machine*. The app encapsulates a dynamic song, aptly described as 'a song that changes just like we all do.' Depending on environmental variables, such as location, weather, and time, the music reshapes itself, offering listeners a sense that the song possesses its unique consciousness.

The primary research questions guiding this phase were:

1 How can semantic technologies assist musicians during the production process?
2 Does the final product align with the composer's aesthetic expectations?

Phase 2 – Mixed-Method Approach: Employing a concurrent qualitative and quantitative design, this phase explored the user's perspective. A pilot user test of the *Semantic Machine* app sought to understand:

1 the ways listeners perceive the contextual changes in the music and
2 whether they see potential in such music formats becoming more widely adopted.

Literature indicates audiences may resonate with contextual listening experiences. However, empirical validation remains scarce, underscoring the importance of this research. Since users had no direct interaction with the project, the study did not need to access the app's usability.

What – Conduct a pilot user study of the *Semantic Machine* app to ascertain how listeners perceive the contextual changes and if they think this type of experience has the potential to become more widely adopted.

Why – The study hopes to get an overview of the necessary variables that might influence participants' interactions with the work and the perceived viability of contextual listening experiences in a music consumption marketplace.

Who – University students, a mix of music students and non-music students.

Where – The study can be done online; a link to the app and the survey was provided.

When – Over one month during the semester.

Research Methods

Participants were invited to experience the app and then complete an online survey. They then entered the draw to win one of two in-ear monitors worth over $500 each.

The study asks participants to listen to the app a minimum of five times, listing their location and time of day for each listen. Each listen took approximately 4.5 min. The total listening time required was 22.5 min over the required five times. There will be a different time and/ or location each time it is logged. It can be the same location at different times, for example. The participant should have a total of five listens within a week. Only a general answer is required for the participant's location, for example, home, work, University, etc. This helped to determine a generalised account of how their listening experience may be affected. For example, if participants just listened to the app in the same location at the same time each day, the listening experience may be very similar. This will affect their overall listening experience, so it was necessary to make an allowance. The app was free to participants via a web link supplied via the survey and participant information sheet. This link opened a web app version of the app in an internet browser, working on all phones.

The survey design considered demographic information, music consumption practices, musical taste and experience, technology, and online behaviours.

This app required location services. Once permitted, the app sends the GPS data to the weather API and Google Maps API. These APIs return data about weather conditions and their location at the given time, which the app then uses to arrange and organise audio playback. Participants' data remains anonymous and is not stored, ensuring privacy.

Participants

The pilot study aimed to have a participant pool 40, divided equally between music and non-music students. The inclusion of questions about participants' previous experience with music ensured an even more refined understanding. As this was a preliminary study, the focus was not on deciphering consumer behaviour but on gaining insights to frame future work in this domain and improve user experience.

Participants were students from the University of Western Australia. All participants were 18 or older. Given the specific nature of this group, results inherently had bias. However, the purpose is not a prediction but an exploration of the potential for larger, more extensive studies.

Unfortunately, only a total of 14 participants participated in the study. This was due to the time of year with Exams and the end of the semester. Unfortunately, the study had a deadline for a conference paper delivery.

Instruments

The project used online survey software. The app was presented as a web application so mobile phone brands would not hinder the experience of the work. The analysis was done in Microsoft Excel.

Results

A notable outcome was that most participants were aware of music changes based on the time of day. However, given the study's demographic and short span, few could identify changes due to location and weather. A few listeners did correlate weather with the music, noting changes that matched the weather's ambience. Feedback from the participants indicated a positive reception, with many noting an enhanced connection to the music, attributing it to the dynamic nature of the compositions. For instance, Respondent 4 mentioned, 'It made me want to keep listening to see how it kept developing over time. The bass, in particular, made me feel really warm.' Another, Respondent 8, valued the personal connection the Transmutable Music allowed, stating, 'I like that everyone can feel like they're listening to a piece of music that's very particular to their individual space and time.'

75% of participants agreed that the music sounded very different each time they heard it.

84% liked that the song sounded different each time it was played.

67% would like to hear music by other artists this way.

67% can see these types of musical experiences becoming more common.

In conclusion, the small pilot study suggests a promising avenue in Transmutable Music, emphasising the need for more artists to explore this innovative concept. The future of Transmutable Music could be promising, especially as more tools emerge to aid its composition. However, with a study this small, the results can only be used to inspire future research.

The *Semantic Machine* is an emblem of the untapped potential of Transmutable Music music, echoing the possibilities of seamlessly marrying interactive technology with art. The app underscores such ventures' viability and ability to captivate audiences. However, it also highlights the pressing need for more research and tools, ensuring this burgeoning musical frontier becomes accessible to a broader range of artists and the public.

Summary

The chapter suggests that understanding audience perception through user studies is crucial for designing Transmutable Music Systems that offer extended musical experiences.

Understanding user experience is essential for engaging and immersing the listener, and this understanding should influence all design stages. This process involves personal approaches and tools extending the listener's experience, drawing from human-computer interaction and music interaction.

User testing should be employed, especially for projects where the audience plays an active role in shaping the music.

It is essential to account for individual differences in perception when crafting user experiences. It is difficult to predict how users will interact with Transmutable Music Systems.

In Transmutable Music, iterative prototyping is a crucial part of the creative process. This approach allows composers and producers to refine their work progressively, ensuring that the music aligns with the dynamic environment of the intended medium.

While artists should trust their vision, flexibility and user testing are vital to achieving a successful Transmutable Music project.

User testing can be integrated into the development of Transmutable Music projects to enhance the end-user experience.

User studies should be based on the project's scope and available resources. Pilot studies are recommended in the first instance.

It is important to clearly define research questions, understand your target audience, and choose the appropriate research method to answer those questions.

The case studies on *One Drop* and the *Semantic Machine* projects provide an overview of how one might approach a user study.

The *One Drop* and *Semantic Machine* case studies illuminate the potential for Transmutable Music to reshape how we interact with and experience music.

Although the *Semantic Machine* pilot study was limited, it indicates that Transmutable Music forms are intriguing for artists to explore.

Continued research and the development of new compositional tools are needed to make these innovative experiences more widespread and accessible.

With technology's constant evolution, the music industry stands on the precipice of a new era where music is not just heard but experienced in tandem with the listener's context. As the lines between technology and artistry blur, the future of music could promise a personalised symphony, ever-changing and ever-evolving.

As well as evaluating the experience of a Transmutable work, in the early stages of development, it is more critical to evaluate experience on the amount of variability the musical playback can provide. Variability is explored in the next chapter in detail.

Exercises

Exercise 4.1 – Musical Design and User Experience

Have a look at the following videos. All these examples would be considered Transmutable Music. Create a diagram of how the projects work within the musical design model. See Chapter 3, Figure 3.7, and Chapter 5, Figures 5.24 and 5.25, for an example of its use. If unsure, think about how you might want to approach it.

Mobile App
https://www.youtube.com/watch?v=f1LuxWeo11w
Sound Walk
https://vimeo.com/24252332
Driving
http://reactifymusic.com/portfolio/vw-underworld-play-the-road/
Flight
https://www.youtube.com/watch?v=zTETqfC6vvA

Exercise 4.2 – Conduct Your Own User Testing

1 Spend some time exploring the following browser-based examples. Conduct your own assessment on how you would rate the user experience. Think about how you might compose the music within this system. How is the system designed? Also, think about the listening experience. Does the music keep you interested? Did you get bored quickly? Did you get frustrated with it? Is it intuitive to use?

- https://www.incredibox.com/demo/
- http://mmorph.massivemusic.com/
- https://reactifymusic.com/static/locus/

2 Find two examples of an interactive-based music project and conduct your own user testing on them.

Exercise 4.3 – User Experience Evaluation and Discussion

1 Go to the *Is this Your World* Max patch on the Companion Website. Once you have downloaded it, play around and explore your own user experience.

　i With this functionality, how would you rate the user experience?
　ii How long would someone want to engage with this idea, and how could it be improved?
　iii Why isn't this an interesting example of an interactive work?
　iv How different can the track sound?

2 Go to the *One Drop* Max patch on the Companion Website. Once you have downloaded it, have a play around and explore your own user experience.

　i With this functionality, how would you rate the user experience?
　ii How long would someone want to engage with this idea, and how could it be improved?
　iii Does the fact that this music was composed for this format improve the experience?
　iv How different can the track sound?

Notes

1 https://www.ninjajamm.com/
2 https://generativemusic.com/reflection.html
3 https://www.arpjournal.com/asarpwp/composing-and-recording-for-fluid-digital-music-forms/
4 https://www.semanticaudio.ac.uk/

References

Babbie, Earl. 1990. *Survey Research Methods.* 2nd Edition. Belmont: Wadsworth Publishing.

Chen, Stephen, Shane Homan, Tracy Redhead, and Richard Vella. 2021. *The Music Export Business: Born Global.* London, United Kingdom: Routledge.

Creswell, John W. 2009. *Research Design Qualitative, Quantitative, and Mixed Methods Approaches.* 3rd Edition. Thousand Oaks, CA: SAGE Publications Inc.

Dawson, Catherine. 2006. *A Practical Guide to Research Methods, a User Friendly Manual for Mastering Research Techniques and Projects.* 2nd Edition. Oxfordshire: How to Books Ltd.

IFPI. 2012. "Digital Music Report 2012." http://www.ifpi.org/content/library/DMR2012.pdf.

Longhurst, B. 2011. "Audience Research." In *Encyclopedia of Consumer Culture*, in D South, 73–75. Thousand Oaks, CA: SAGE Publications, Inc. 10.4135/9781412994248.n28.

Metaswipes. 2023. "Mobile Gestures: Fascinating History & Facts." https://Metaswipes.Com/. 2023.

Redhead, Tracy. 2015. "Composing and Recording for Fluid Digital Music Forms." *Journal on the Art of Record Production* Proceeding (Issue 10). http://arpjournal.com/composing-and-recording-for-fluid-digital-music-forms/.

Redhead, Tracy. 2020. *Dynamic Music: The Implications of Interactive Technologies on Popular Music Making*. NSW, Australia: University of Newcastle.

Woodward, Ian. 2011. "Methodologies for Studying Consumer Culture." In *Encyclopedia of Consumer Culture*, in D. South, 1664. Thousand Oaks, CA: SAGE Publications, Inc. 10.4135/9781412994248.

Zuboff, Shoshana. 2019. *The Age of Surveillance Capitalism*. London, England: Profile Books.

5 Variability

Variability and Variation

Given that Transmutable Music is fluid and changeable, some amount of musical variability must be composed to offer something more than a fixed or static piece. Variability within the playback is one way to expand this experience.

It is easy to get confused by the term's variation and variability. They are similar but have different meanings in this context. Variation is a well-established musical term. To add to this confusion, using variation as a composition tool will create more variability in your work. Variation is defined as making changes to aspects of a work's structure or components. It introduces change while maintaining a connection to the original composition. Music has long celebrated the concept of variation, as can be heard in the classical form, theme and variations, or when a jazz soloist improvises on aspects of a melody.

The essence of musical variation, as discussed by Nelson in 1948, lies in the artful balance between alteration and preservation. To achieve variation, a piece of music must retain identifiable elements of the original while introducing changes in harmony, structure, rhythm, timbre, melody, figuration, or expression. This interplay ensures that the piece remains a variation of the original rather than becoming an entirely new piece. True variation maintains a thread of continuity; without it, we havecontrast, where the connection to the original theme may no longer be perceptible (Nelson, 1948).

Variability, however, refers to the state of being variable, changeable, adaptable, fluid, malleable, or flexible. So, variability could be understood as a measurement of how much variation is composed in a work. Within Transmutable Music, variability can be measured across all components, including the control system, musical architecture, and content. This chapter will provide practical approaches to increasing variability in Transmutable Music works. Each of these approaches will include case studies or tutorials.

DOI: 10.4324/9781003273554-7

Designing Variability in a Transmutable Music System

'Processes throw up momentary configurations which have no sooner happened than they are past: the experimental composer is interested not in the uniqueness of permanence but in the uniqueness of the moment' (Nyman, 2000, 9).

Transmutable Music encapsulates the transient nature of musical arrangements, echoing Nyman's perspective that the value lies in the fleeting uniqueness of each moment, not in its lasting existence. This aligns with the idea that the musical output of a Transmutable Music system is a singular occurrence amongst a multitude of potential combinations, giving it a distinct shape within the momentary present.

Contrary to the traditional notion of fixed musical structures, Transmutable Music presents a fluid form, constantly reshaped by its inherent processes. These processes generate multiple, distinct forms; each experienced as a transient, unique moment by the listener. The listener's engagement with the music is confined to the immediate rendition, making Transmutable Music intrinsically tied to the experience of 'now.'

Transmutable Music extends the listening experience in two principal ways: through feedback mechanisms that signal potential changes in the music and through the inherent variability of each playback, ensuring a unique encounter every time. Assessing the feedback aspect of Transmutable Music necessitates user studies to gauge audience perception and comprehension (as discussed in Chapter 4). However, during the initial phases of development, the potential for diverse experiences can be determined by the variability offered in playback.

Variability can be used to extend the experience of Transmutable Music so that the music differs with each playback. This can be achieved by designing variability within all components of a Transmutable Music system.

- *Variability in the Musical Architecture* – the following compositional tools can be used to extend musical options, including variation, contrast, branching of sections or phrases, layering, blending and transitions.
- *Variability in the Content* – Composing more musical and sound content options, increasing sound design possibilities with digital instruments, patches, sound library or procedural audio.
- *Variability in the Control System* – designing the control system to control the variability options designed in the content and musical architecture.

The interplay between these elements of Transmutable Music and the design of variability is pivotal. As shown in Figure 5.1, the control system meticulously orchestrates the variability in content and structure to yield the dynamic outcome of the musical output.

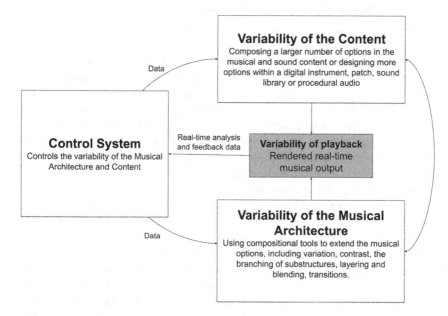

Figure 5.1 Variability within the Musical Design Model for a Transmutable Music System.

The tools and methods to design variability in the musical framework of Transmutable Music systems are boundless, ranging from simple to highly complex.

For instance, straightforward variability might involve tempo, pitch, or volume changes, and even the mix parameters of the piece. This could entail altering the spatial characteristics of the instruments and the layers or incorporating dynamic elements such as compression or equalisation. Advanced techniques might encompass algorithms and randomised parameters affecting pitch, structure, tempo, or rhythm.

Producing more variability will help to produce a better experience. Aspects where variability could be introduced are pitch, timbre, metre, rhythm, tempo, time signature, dynamics or energy, FXs and DSP, instrumentation, melodies, harmony (key, mode, scale, chords), section or phrase structure, mixing parameters, and form.

Composing variability can create variation and contrast in the musical content at different hierarchical levels. Variability on the Master level, for instance, would be considered the most straightforward approach. In contrast, changes to the sound object or lower hierarchical levels would be seen as complex.

Variability can also be achieved through the design of Transmutability see Chapter 6. I present evaluation criteria in this section's final Chapter 7 because evaluating a Transmutable music work requires elements of variability, transmutability and experience.

Tools and Approaches for Composing Variability

Composing variability introduces a unique challenge and perspective to musical creation. For traditional compositions, the goal is to find the perfect part that flawlessly fits the structural puzzle. Transmutable Music, however, demands a broader vision where each musical segment is conceived with multiple potential variations. This necessitates a multi-dimensional approach to composition, contemplating how various versions of a melody, for instance, can coexist and function within every other concurrently playing element. Based on the structural framework you establish, it is an intricate web of connections, a 360-degree compositional outlook where each layer can influence and potentially alter every other. In Chapter 3, I introduced a model for composing Transmutable Music; now, we will look at the practicalities of crafting variability.

Two primary methods can be employed to design variability within a Transmutable Music system: by composing an extensive collection of materials and techniques or by a minimalist strategy, applying variability to a more limited set of materials through transmutability. This design approach allows a minimalist array of materials to yield a rich spectrum of musical expressions.

As a composer, I am always in pursuit of variability. When I create a new instrument or a piece of music, I aim to embed it with diverse sonic possibilities. For example, in my approach to sound design, I might assemble an instrument rack in Ableton Live with three instruments blended together, each with its own audio effects (FX). The sound's evolution over time is revealed by manipulating parameters and exploring intrinsic qualities, such as modulating a wavetable synth's oscillator position with a slow LFO.

This method steers away from the quest for a single 'perfect' sound, focusing instead on a soundscape rich with variety. As melodies take shape, I consider how they might be transmuted across different instruments to enhance their variability. With harmonies, I explore chord progressions that offer variation or contrast, which fuels the organic growth of the music. New layers and modifications continually inspire further variability.

The outcome is a musical structure or system with numerous layers, each capable of variation when played alone or together. This evolving process of building, modifying, and experimenting enriches the composition and invigorates the creative journey. While this approach suits layering (vertical orchestration) well, it applies equally to branching (horizontal resequencing).

Even straightforward tempo, pitch, volume, or DSP changes can inject variability into a piece. The mix itself can become a canvas for variability, with spatial effects, volume modulation, and dynamic processing like compression or EQ, alongside the creative application of audio and MIDI effects. Randomising parameters like pitch, structure, tempo, and rhythm can produce unique musical outcomes.

For a more organised and computationally efficient approach, MIDI and coding offer a broad spectrum for introducing variability. These tools allow for nuanced sweeping changes, including tempo shifts, harmonic adaptations, and instrumental rearrangements. Layering and structuring MIDI sequences and coded programs result in complex yet efficient arrangements.

Moving into higher complexity, melody and harmony modification algorithms come into play, alongside structuring compositions in layered, accumulative forms or open frameworks that adapt variably. Such advanced methods deepen the interactive and dynamic facets of the listening experience.

Variation

While variability refers to the capacity for change, the technique of variation is an essential tool used to achieve variability. As discussed earlier in the chapter, variation can be used across the entire structure of the content and architecture. It involves the alteration of original aspects to produce a change. This could include melodic, harmonic, and rhythmic variations. For true variation, (Nelson, 1948) outlines that certain musical elements must remain constant while others change (Nelson, 1948). When all elements change, this could be understood as contrast. Contrast is essential to use in a branching (horizontal resequencing) approach.

Melodic Variation Techniques

Melodic variation is a pivotal technique for enriching and expanding your musical creations. Beyond these hands-on techniques, melodies can also be dynamically altered through data-driven changes within a system. For instance, in video game composition, melodies can evolve in real-time using various software-enabled techniques.

Sweet (2015) provides a comprehensive catalogue of strategies to minimise repetition and refresh the melodic line. These include:

1 Transposition: Shifting the melody's pitch up or down by a set interval.
2 Reharmonisation: Modifying the harmonic backdrop, which, in turn, changes the melodic context.
3 Inversion: Keeping the initial pitch but reversing the direction of the subsequent intervals.
4 Retrograde: Reversing the sequence of the melody.
5 Permutation: Rearranging the order of the notes.
6 Rhythmic Displacement: Shifting the melody in time, either forward or backward, by a specific rhythmic unit.
7 Truncation: Shortening the melody.
8 Expansion: Lengthening the melody.
9 Rhythmic Alteration: Changing the rhythm of the melody without affecting the pitches.
10 Melodic Alteration: Changing the pitches without altering the rhythm.
11 Thinning: Stripping away specific notes from the melody.
12 Ornamentation: Embellishing the melody with additional notes (Sweet, 2015).

You could establish a set of these techniques or 'rules' to dictate how the melody should transform. Working with MIDI data simplifies this process, as it demands less computational power than audio files. See the Chance and MIDI Tutorial: Utilising Randomness in Max for Live in Chapter 6 for details on achieving this. Nonetheless, you could also craft melodic variations in audio form, selecting variations randomly or allowing the system to control these changes via MIDI or another music representation model. Details on implementing these changes are explored in Chapter 6 on Transmutability.

Another avenue for incorporating melodic variation is using embellishments or fills. Much like a drum part is kept engaging with fill-ins, your melodic phrases can be punctuated with occasional short fills or embellishments, ensuring the music remains dynamic and ever-changing. The following tutorial looks at Harmonic variation.

Tutorial 5.1 – Generative Work Using Variation

This is a straightforward exercise for creating variations to an 8-bar loop. We can do this by adding, subtracting, and changing parts, notes, instrumentation and note lengths. Based on this 8-bar loop, we will create a generative ambient track that randomly moves through loops and changes FX and panning parameters. You can do the tutorial by following the instructions or create your own work. You can download the finished Ableton set from the Companion Website. This is an excellent way to get an overview of the tutorial before you start. It can also help if you get stuck. Check the Companion site for videos, links and help if this tutorial is difficult.

EXERCISE 1 – HARMONIC VARIATION

1 Open a new Ableton Live project.
2 Add a MIDI instrument of your choice to a MIDI track. I used the Grand Piano instrument to keep it simple.
3 Compose 8 bars with a four-chord harmonic progression. I used i VII VI VII7 in A Minor. The chords are 2 bars long and are Am G F G7. Rename the clip or loop to 'original.' See Figure 5.2.
4 For our first variation, let us change two of the chords. I changed the second chord to a ii* or B dim and the third chord to the v or Emin. So this variation is Am Bdim Em(second inversion) and Gmaj7. Rename to 'Vchords.' See Figure 5.3.
5 Next, duplicate the first loop we made, then remove the third interval of the chord to create some nice power chords or 5ths. Rename the clip to 'V5ths.'
6 Now, remove one or two of the chords. I removed the second and fourth chords. Rename 'Vsub5ths.'

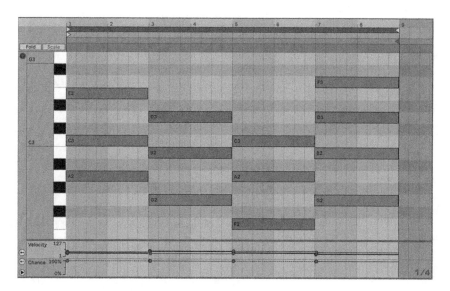

Figure 5.2 Screenshot of *Ableton Live.*

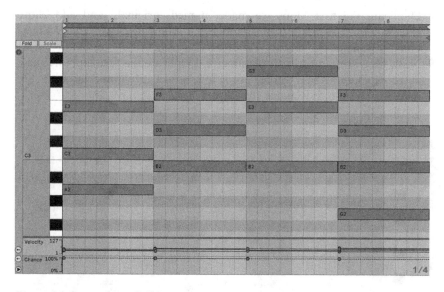

Figure 5.3 Screenshot of *Ableton Live.*

7 Duplicate the loop we made in task 5 named 'V5ths,' and now add the 3rd intervals back; however, change two chords to the major scale. I have changed the progression by adding a maj 3rd interval to the first chord and a min 3rd to the second chord. The progression becomes I vii VI VII7, A Gm F Gmaj7. Rename to 'Vkey.' See Figure 5.4.

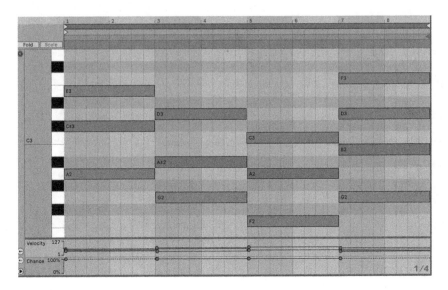

Figure 5.4 Screenshot of *Ableton Live*.

8 Duplicate the first loops we made, named 'original,' and add chord extensions to the progression. I added the following: Am7 Gmaj9 Fmaj7 Gmaj9. Rename the clip to 'Vchord Ex.'

9 Next, we will change the rhythm/metre. Duplicate the first clip, 'original,' and create a series of 8th-note chords to produce a rhythmic variation. See mine in Figure 5.5. Rename this clip 'V8ths.'

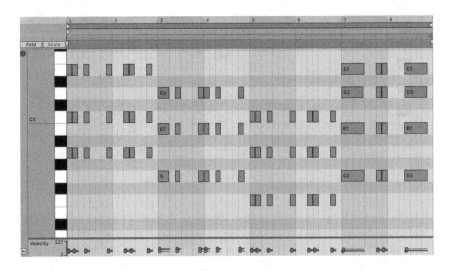

Figure 5.5 Screenshot of *Ableton Live*.

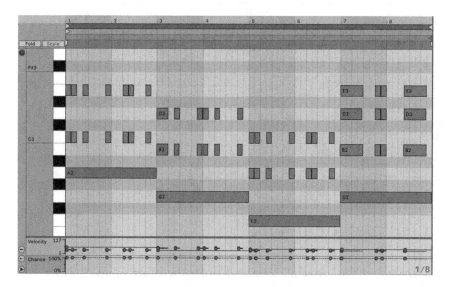

Figure 5.6 Screenshot of *Ableton Live*.

10 We can do a slight variation on instruction 9, 'V8ths,' by making the root note of each chord go over 2 bars. Rename it to 'V8thnote.' See Figure 5.6.
11 Let us now make some silent loops. I will make three at different lengths so the clips can go out of phase as they are played randomly. I made the following midi clip: 4 bars, 8 bars, and 12 bars. Then rename the clips 'Vsil4b' Vsil8b' and Vsil12b.' See Figure 5.7.

Figure 5.7 Screenshot of *Ableton Live*.

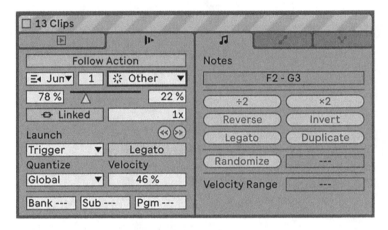

Figure 5.8 Screenshot of *Ableton Live*.

12 We will organise our first instrument variation loops. Create three versions of the 'original' loop and order the other loops however you like. I have created the following order; see Figure 5.7.

13 Next, let us set up the clips to keep playing sequentially. Select all the clips by holding down shift. Now go to the clip view, then the second tab with the follow-through info. Click on the 'Follow Action' button so it is yellow. Then select the first drop-down menu, 'jump to' and the second to 'other.' This will set up the clips to play randomly. So, when a clip finishes, another will start. Finally, change the launch drop-down menu to Trigger. See Figure 5.8.

14 You can also randomise the velocity of the note (or how loudly it is played back each time). I have made this 46% randomised.

15 Add an EQ eight onto the device window. Add a high pass filter at around 79 Hz.

16 To add sound design, drag an Auto Filter onto the track's device window.

17 Next, drag an LFO onto the same track.

18 Move the Freq to around 1KHz on the Auto Filter and increase the res dial to 61%.

19 Map the LFO Map button to the Freq of the Auto Filter, then reduce the rate to 0.03 Hz, the depth to 81% and the Offset to 40%. See Figure 5.9.

Figure 5.9 Screenshot of *Ableton Live*.

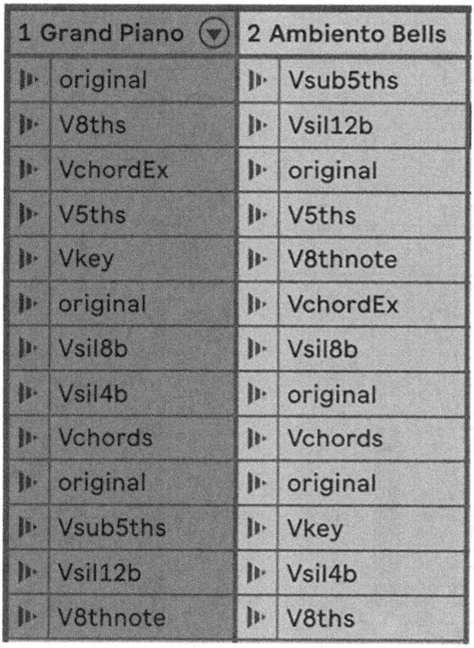

1 Grand Piano ⦿	2 Ambiento Bells
▶ original	▶ Vsub5ths
▶ V8ths	▶ Vsil12b
▶ VchordEx	▶ original
▶ V5ths	▶ V5ths
▶ Vkey	▶ V8thnote
▶ original	▶ VchordEx
▶ Vsil8b	▶ Vsil8b
▶ Vsil4b	▶ original
▶ Vchords	▶ Vchords
▶ original	▶ original
▶ Vsub5ths	▶ Vkey
▶ Vsil12b	▶ Vsil4b
▶ V8thnote	▶ V8ths

Figure 5.10 Screenshot of *Ableton Live*.

EXERCISE 2 – INSTRUMENT VARIATION

1 Duplicate the grand piano midi track and rearrange the order of the clips. You can see mine in Figure 5.10.
2 Let us change the instrumentation to layer the harmony; you can use whatever instrument you like. I have used the 'Ambiento Bells' Instruments located in Instruments/Wavetable/Synth Keys/. See Figure 5.10.
3 Next, you can change the colour by right-clicking on the track and selecting a new colour, then right-clicking again and selecting 'assign track colour to clips.' See Figure 5.10.
4 Now we will update the reverb return; I have used the ambience preset reverb Audio Effects/Reverb and Resonators/Reverb/Room/Ambience. Drag the Ambience reverb into the reverb return.
5 Update the delay in the delay return so that the ping pong button is selected, then add some sends from both MIDI tracks. See Figure 5.11.
6 Adjust the gain on your tracks so that the master track is peaking at around -15db.
7 Finally, pan the Grand Piano track to the right and the Ambiento Bells to the left.
8 To add some timbre variation to the Ambiento Bells, turn on the Osc 2 tab.

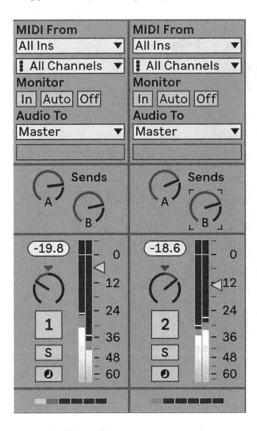

Figure 5.11 Screenshot of *Ableton Live.*

9 Now drag a shaper device into the track after the wavetable preset. Set up a peak triangle with a rate of 0.01 Hz and a depth of 78%. Now press the map button on the top left and map it to the gain on the Osc 2 (Figure 5.12).

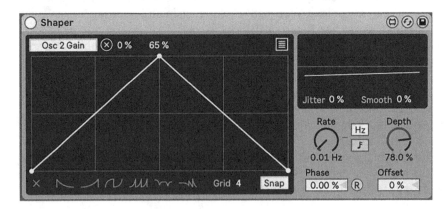

Figure 5.12 Screenshot of *Ableton Live.*

EXERCISE 3 – ADDING SOUND VARIATION

1 Add the Sunrise Wave preset into a MIDI track. It is located in Instruments/Wavetable/Ambient and Evolving/Sunrise Wave.
2 Create a new MIDI clip 8 bars long and add the note A3 lasting the whole 8 bars.
3 Adjust the velocity to 64 at the bottom left.
4 In the Clip view, change the velocity range to +33. See Figure 5.13.
5 At the bottom of the clip, click the small round button next to velocity. Make the chance 64%. See Figure 5.13.
6 Rename the clip 'A3' and duplicate it.
7 Rename the new clip 'A1' and move the 8-bar MIDI note down two octaves (Shortcut – select the MIDI note, then hold shift and press the arrow to lower the MIDI note by an octave.)
8 Duplicate the A1 clip and move the MIDI note up to 'A2.'

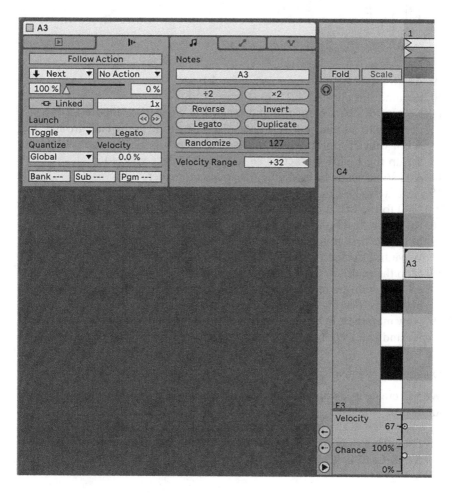

Figure 5.13 Screenshot of *Ableton Live.*

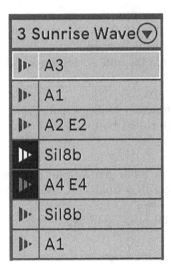

Figure 5.14 Screenshot of *Ableton Live.*

  9 Add a 4-bar E2 note from bar 5 for 4 bars.
10 Rename this clip 'A2 E2.' See Figure 5.14.
11 Next, add 8 bars of silence and rename it 'Sil8b.' See Figure 5.14.
12 Feel free to add more MIDI notes that can drone in the background, like A4 or other E notes, which may sound nice, too. See Figure 5.14.
13 Set up the Follow-on action so random clips are selected.
14 Add some sends to help glue the sounds together.
15 Next, we will use modulation to create a constantly moving sound and add an LFO to the MIDI device chain: Audio Effects/Modulation/LFO.
16 Add a multi-map M4L device after the LFO. See Figure 5.15.
17 Map the first multi-map parameter to the volume of track 3 (Sunrise Wave track). See Figure 5.15.
18 Map the multi-map M4L device to each of the tracks and send dials. Play around with the min, max and curve parameters to get slightly different mappings. See Figure 5.15.
19 Now map the LFO to the control of the multimap; change the rate to 0.04 Hz and the Depth to 54%. See Figure 5.15.

EXERCISE 4 – AMBIENCE

We will create a custom instrument with macro effects to add another layer of ambience.

  1 Drag the Echo Bay preset into a new MIDI track Instruments/Wavetable/ Synth Lead/Echo Bay.

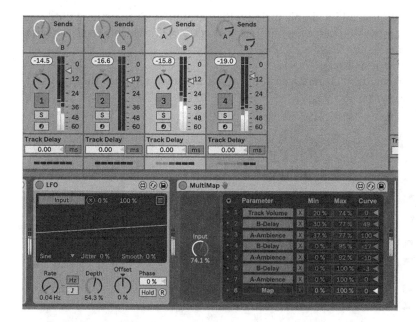

Figure 5.15 Screenshot of *Ableton Live.*

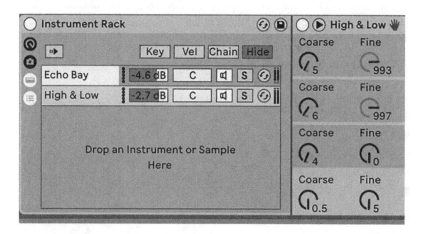

Figure 5.16 Screenshot of *Ableton Live.*

2 Turn on the second Osc in the tab.
3 Right mouse click on the top of the device and select Group (or click on the top bar of the device and press command G (Mac) or control G (PC)) to create an instrument rack.
4 Next, drag the High & Low operator preset in Instruments/Operator/ Ambient and evolving/High & Low. See Figure 5.15.
5 Adjust the volumes to blend the sounds together, as shown in Figure 5.16.

6 We will now add a simple melody. Play around with something until you are happy. You can view my one in Figure 5.17, which ended up being 20 bars.
7 Rename it as Mel 1.
8 I duplicated Mel1, removed some notes, and then renamed the clip Mel2. See Figure 5.18.

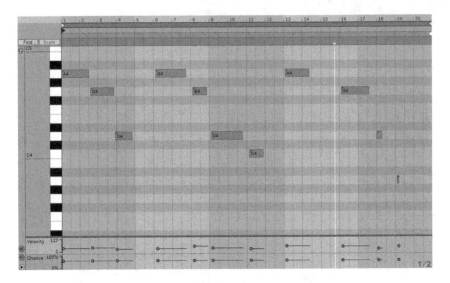

Figure 5.17 Screenshot of *Ableton Live.*

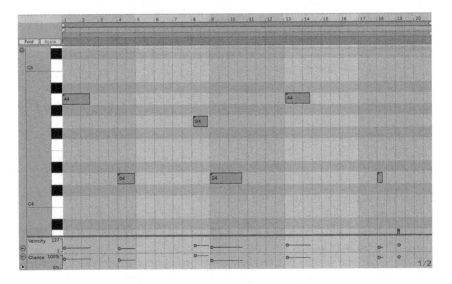

Figure 5.18 Screenshot of *Ableton Live.*

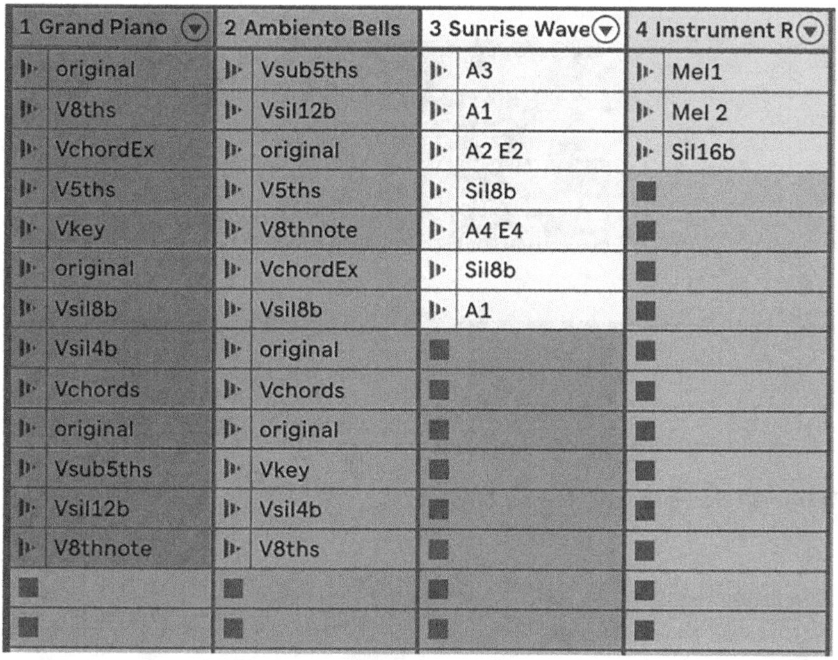

1 Grand Piano ⓥ	2 Ambiento Bells	3 Sunrise Wave ⓥ	4 Instrument R ⓥ
⃰ original	⃰ Vsub5ths	⃰ A3	⃰ Mel1
⃰ V8ths	⃰ Vsil12b	⃰ A1	⃰ Mel 2
⃰ VchordEx	⃰ original	⃰ A2 E2	⃰ Sil16b
⃰ V5ths	⃰ V5ths	⃰ Sil8b	◼
⃰ Vkey	⃰ V8thnote	⃰ A4 E4	◼
⃰ original	⃰ VchordEx	⃰ Sil8b	◼
⃰ Vsil8b	⃰ Vsil8b	⃰ A1	◼
⃰ Vsil4b	⃰ original	◼	◼
⃰ Vchords	⃰ Vchords	◼	◼
⃰ original	⃰ original	◼	◼
⃰ Vsub5ths	⃰ Vkey	◼	◼
⃰ Vsil12b	⃰ Vsil4b	◼	◼
⃰ V8thnote	⃰ V8ths	◼	◼
◼	◼	◼	◼
◼	◼	◼	◼

Figure 5.19 Screenshot of *Ableton Live*.

9 Add as many melodic variations as you like, then create the 'follow-on' commands per exercises 1 and 2. I just added 16 bars of silence. See Figure 5.19.

10 Next, we will add some effects to add more variation. Click on the High & Low instrument in the instrument rack. First, map the filter in the High & Low Operator device to Macro 1. You can do this by right mouse clicking on the dial and selecting Macro 1. See Figure 5.20.

11 Add a Spectral time delay preset called 'Delay Spray.' Audio Effects/ Delay & Loop/Spectral Time/Delay Spray. Map the dry/wet dial to Macro 2 by clicking the right mouse on the dial. See Figure 5.20.

12 Next, add the Diffused Long Cascades after the Delay Spray but inside the High & Low rack. Audio Effects/Delay & Loop/Echo/Ambient Spaces/Diffused Long Cascades. Map the Wet dry to Macro 3. See Figure 5.20.

13 Next, we will map the Echo Bay instrument. Map the gain of Osc 2 to Macro 4.

14 Then, add the Ethereal Canyon device in the browser in Audio Effects/ Delay & Loop/Echo/Ambient Spaces/Ethereal Canyon. Map the dry/wet to Macro 5.

(a)

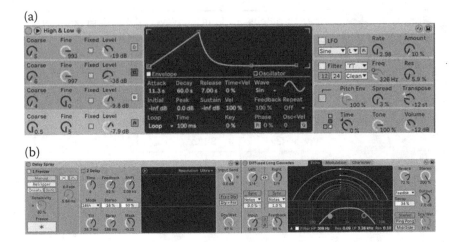

(b)

Figure 5.20 Screenshot of *Ableton Live.*

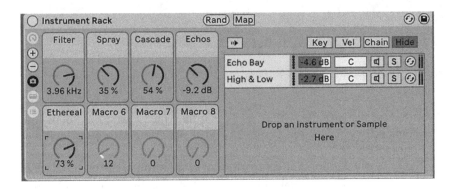

Figure 5.21 Screenshot of *Ableton Live.*

15 Now, Rename the Macros so you can remember what they do. See Figure 5.21.
16 Lower the gain of the track to about -11db. Then, add a Compressor Device after the instrument rack. Lower the threshold to around -50 and increase the make-up gain to around 9db. To get a softer tone.
17 You can now use LFOs to Map these changes to evolve as the track plays. See Figure 5.22.

EXERCISE 5 – TWEAKING

1 Add an arpeggiator before the Grand Piano instrument on track one. It is in MIDI effects.

(a)

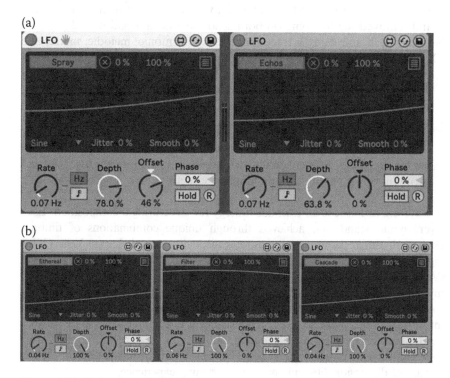

(b)

Figure 5.22 Screenshot of *Ableton Live*.

Figure 5.23 Screenshot of *Ableton Live*.

2 Play around with the settings and find something you like. I set the style to Random with four steps. I also transposed to Am. See Figure 5.23.
3 Next, add a random midi effect; I set the chance to 87% with three choices and four scales.

4 Finally, I added a Wanderer filter delay onto the end of the device chain and removed the mapping on both sends so I always had max delay.

Now, sit back and enjoy a track made using harmonic, melodic, metre, and FX parameter variation. It is based on a very simple 8-bar chord progression. Keep tweaking it until you are happy with it. You can download the finished Ableton Live project from the Companion website.

Layering, Blending and Vertical Orchestration

'Layering is like using colours to distinguish one thing from another (for example, using different coloured labels for identifications), whereas blending is like mixing colours to make something new' (Vella, 2003, 114).

Layering in music is an essential strategy for structuring a composition. Each layer should stand out, achieved through unique combinations of timbres, pitches, rhythmic patterns, volumes, and distinctive sound shapes. Vella (2003) highlights our ability to perceive these layers as if they occupy physical space (Vella, 2003). Imagine being at a live concert: you can visually locate each musician – the drummer, guitarist, and bassist – on stage. Similarly, our brains spatially locate sounds when we listen to music. In stereo mixing, typical for most music, the sound engineer places each instrument in a unique sonic space between headphones or speakers using panning, allowing us to discern and enjoy each sound as if it were positioned on a stage before us. This spatial 'placement' creates a sense of dimensionality, enhancing our listening experience.

To build a rich musical texture, the process of layering involves the artful combination of melodic, harmonic, rhythmic elements, or even noise. This technique is not only vital for composition but also serves as a means to introduce variability in Transmutable Music.

Even subtle changes, such as tweaking the volume of layered tracks or shifting the panning to alter the perceived origin of a sound, can significantly influence the emotions and energy within a piece of music. These adjustments can add depth and variety, making each listening experience unique.

When discussing instrumentation, we refer to the selection of instruments used in an arrangement. Variability here could mean changing the instrument's timbre or substituting one instrument for another. For instance, replacing a horn section with strings can significantly alter the sound character of a piece.

In game audio, this concept of layering is often referred to as 'vertical' changes for example 'vertical orchestration' or 'vertical remixing.' The term 'vertical' is used because tracks and instruments are listed vertically in a DAW or on sheet music. Sweet (2015) details 'vertical remixing' as an interactive method where layers are manipulated to adjust the intensity and emotion of the music, a technique we will demonstrate in Tutorial 12.1 in Chapter 12.

Blending, in contrast to layering, is where the individual sounds merge to such an extent that their distinctiveness is lost, resulting in a new, fused sound. Robert Erickson (1975) depicted blending as a delicate balance where individual instrument sounds are no longer distinguishable, and the fused sound consistently commands attention (Erickson, 1975).

In my work, I have explored how layers can interact in a way that allows them to blend seamlessly when played together in any combination and how they can create intriguing effects when combined with other musical elements. This approach is not just about stacking sounds but also about how these sounds interact to create dynamic shifts within the music.

In order to explore these concepts in more depth, I will present a case study of the *One Drop* composition (referenced in Chapter 4).

Case Study: Composing Variability through Layering, Blending, and Orchestration (Master Level)

The Ableton project, *One Drop*, and the stems are provided on the Companion Website.

INTRODUCTION

I encountered many challenges in my early research (from 2010) on composing fluid music forms. I learned valuable lessons, especially in creating *One Drop*, a composition with 18 amorphous stems; see Figure 5.24 for the musical design of the work. This case study reflects on these experiences, offering insights into the process of composing music that is not only layered but also responsive to listener interaction.

One Drop - Musical Design

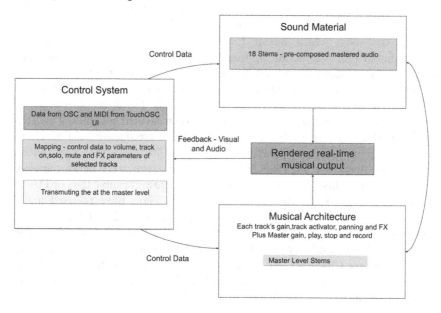

Figure 5.24 Musical Design Model for *One Drop*.

STAGE 1: EXPLORATORY PHASE AND ALPHA PROTOTYPE

My initial venture into this field involved examining interactive music applications, which were limited then; however, the BBC 2004 'Fat Boy Slim' remix game was a highlight (Fat-Boy-Slim, 2004). This led to the development of an alpha prototype interface. This used stems from *Is This Your World?*, which was a part of my album *Walking Home a Different Way* released in 2008. This stage highlighted the first major stepping stone of my journey. The stems were composed for a fixed form exported from the final mix[1] of the track down to 7 stems instead of one master. Even without user testing, the prototype revealed limitations in interface functionality and the variability that could be achieved. This restricted user creativity, resulting in a very limited user experience – a valuable insight that prompted a reimagining of stem recording for enhanced user engagement. An example of this work can be downloaded from the Companion Website.

STAGE 2: EXPERIMENTATION WITH 'FALLING DREAM'

Informed by the insights from the alpha prototype, I wrote and recorded a new song, 'Falling Dream,' to be inherently modular and adaptable. I attempted to record three stylistically different versions of the song to mix into one, but this highlighted several issues: inconsistencies in song structure, key changes, and timing inaccuracies inherent to non-quantised, human-played recordings, all contributing to a lack of cohesion. Unfortunately, I was unaware of the concept of Vertical orchestration at the time.

STAGE 3: THE COMPOSITION OF *ONE DROP*

Finally, I chose my song *One Drop* for its inherent adaptability and potential for cross-genre appeal. I had no solid artistic vision for the production and style of the song, which meant I was more willing to experiment with compositional and production approaches. I composed a series of independently complete stems that could harmonise in any combination. Guided by previous challenges, I arranged the guitar and string parts to ensure harmonic richness and versatility across different arrangements. The drum parts required precision; they needed to align rhythmically when layered together. This step underscored the necessity for each instrumental arrangement to work in isolation and unison. There was a total of over 90 tracks that were mixed down to 18 stems, see Figure 5.24. At the time, stems were being released, along with radio edits,[2] by many artists to encourage and enable fans or DJs to remix their songs. Trent Reznor and Nine Inch Nails have released many albums and singles since 2005. www.remix.nin.com (Nails, 2013). Other artists who have released stems include R.E.M Field, Radiohead, Kylie Minogue (SoundCloud, n.d.) and Beyonce (Pandora, n.d.). For more information on the Stem format, see Chapter 8. An Ableton set with the final stems for this track can be downloaded from the Companion Website. A simplified version of the prototype in Max is also available.

ORCHESTRATION AND BLENDING INSIGHTS

The orchestration process demanded a modular approach, where each part was designed to fit seamlessly within any stem combination. The transparent nature of the fluid format necessitates impeccable recording quality (as opposed to the sound of crickets in the background of the lead vocal), as the format would expose every nuance. Three synth stems were crafted from 13 different synth tracks to provide different genre mixes while maintaining cohesiveness. The guitar and string arrangements were strategically simplified to suit the interface design without sacrificing the overall aesthetic. Throughout the compositional experience, I learned that some songs you know as a composer how you want them to sound. Others like this have so many arrangements and production ideas. It was exciting to explore the possibilities of a song with many forms.

VOCAL ARRANGEMENT CONSIDERATIONS

Vocal arrangements posed unique timing and performance challenges. Inspired by the duet-style vocals in *Empire Duet* by Rhythm and Sound (Sound, 2003), I created vocal tracks that could stand alone or blend harmoniously. The emotional delivery of vocals had to be balanced with the rhythmic precision required for the format. Also, we mixed the vocals so they would all work together. This meant the lead vocal was louder than the other three vocal parts. While it was not the best outcome, it was the only way to mix it in all combinations.

FINAL MIXING AND INTERFACE DESIGN

In the concluding phase, I worked with Prof. Julian Knowles to mix the stems, ensuring each stem's volume and EQ settings allowed for a coherent sound in any combination. This stage aimed to provide a high-quality sound regardless of the listeners' mix choices. The user interface (see Figures 4.2 and 4.3) was designed to be intuitive to accommodate users at different skill levels and promote interaction through straightforward controls and effects.

CONCLUSION: REFLECTIONS AND FUTURE DIRECTIONS

This case study illustrates the critical need for adaptability and creativity in composing music for interactive formats. The experiences with *One Drop* highlight the importance of a flexible approach to composition, recording, and mixing to facilitate a transmutable listening experience. As composers, we are tasked with not only creating music but also crafting engaging auditory environments that invite listener participation and reinterpretation. This project was produced many years ago, and while it is difficult for me to listen to it, hopefully, it is educational.

Branching and Horizontal Resequencing

Designing for variation becomes increasingly intricate when you build from the ground up. In this process, the song's structure is not static but fluid, shaped by how its components – or sub-structures – are configured. This is where concepts like branching, crossfading, and smooth transitions come into play. Branching, a technique commonly used in game audio, is known as horizontal resequencing referring to events occurring linearly. It is a way of assembling the music in real time, responding to the actions within the game – for instance, triggering a new musical sequence when a character enters a different environment.

In popular music, this could be using a cue to shift from a verse to a chorus or an entirely different version of a chorus. By designing various permutations of song sections, you introduce a high degree of variability. Consider a basic song structure consisting of two sections, A and B. You can branch this into multiple patterns: ABAB, AABA, BABA, etc. If each section, A and B, has six different variations – A1 through A6 and B1 through B6 – you can create many unique compositions.

For example, combining different section patterns with their variations could yield sequences like A2B3A1B6 or A5A2B5A1, showcasing just a fraction of the potential diversity. With just these variations, you have over 20,000 possible combinations, vastly expanding the piece's variability.

Ensuring seamless transitions between sections is crucial when developing such a branching structure. This could be done through crossfading, where one section blends into the next, or by crafting transitional elements to introduce key or tempo changes. Effective transitions can be programmed directly within the sound material or the control system managing the piece.

Audio middleware tools are particularly adept at facilitating smooth transitions. A composer might choose crossfades that sync with the beat or bar or opt for longer transitional cues that offer a more natural bridge between distinct musical segments. Tutorials 12.4 and 12.5 provide examples of how to achieve this using FMOD.

For compositions where a melodic line needs to resolve before moving on, 'Branching Scores' are also recommended. This method allows one musical phrase to finish cleanly before the next begins, without the overlap of crossfading. Such techniques are versatile and can be adapted to suit various musical styles.

Sweet (2015) outlines three primary methods for horizontal resequencing in game music: crossfading, transitions, and branching. The choice among these methods will largely depend on the structure and nature of the music you have composed (Sweet, 2015).

Let us consider these methods in the context of traditional song structure.

Branching scores act like a relay race: one section passes the baton to the next, with no overlap. This method is straightforward – if your song's structure is AB AB, Section B will not begin until Section A has completely concluded. This method is suitable when you want a clear completion of one musical phrase before the next begins.

Crossfading is another transition technique where one musical section fades out as the next fades in. This approach can be more straightforward than branching and can be executed at any moment, depending on your composition's structure. Elements like tempo and key must be compatible to ensure a smooth transition, allowing the sections to blend seamlessly.

In gaming, music must adapt instantaneously to the player's environment without causing a jarring interruption. For instance, if a peaceful forest exploration suddenly turns into a boss battle, the music must immediately transition from the forest theme to the battle theme. As discussed in Chapter 4 on user experience, players must perceive this musical shift as a direct response to their in-game actions.

Using traditional DAW software for such instant transitions might be challenging. In the upcoming game audio tutorials in Chapter 12, we will use audio middleware to craft transitions that grant you the flexibility to switch between varying musical states, regardless of differences in tempo or key.

One common transition technique involves 'stingers,' such as a crash or timpani roll, to indicate a significant in-game event, such as a point score, and to create a smooth transition between musical themes. Alternatively, a musical bridge or interlude can serve as a connector between contrasting sections, much like a pre-chorus that both builds tension and seamlessly links a verse to a chorus. This is particularly useful if the verse and chorus do not naturally flow into one another. The pre-chorus then serves the dual purpose of heightening anticipation and ensuring musical cohesion.

When composing adaptive music, it is important to recognise that horizontal resequencing and vertical orchestration are not mutually exclusive. Typically, both techniques are combined to create a dynamic and responsive musical score. The challenge arises when working within a Digital Audio Workstation (DAW) without additional programming tools like Max for Live (M4L), which can facilitate precise control over musical events.

For instance, if you are playing a track in a DAW such as Ableton Live and wish to execute a transition on the timeline while simultaneously crossfading tracks, managing this directly within the DAW can be complex. However, this can be achieved by utilising multiple tracks in Ableton's Session View or by programming a custom control device in M4L that orchestrates events and their musical outcomes, like triggering transitions and managing crossfades in real-time.

To practically develop your skills and approaches for composing branching scores and transitions, see the Chapter 12 Adaptive Audio Tutorial Series.

Case Study: Composing Variability through Layering and Branching
(Section Level)

This case study explores the concepts of vertical orchestration and horizontal remixing in a compositional model that integrates a Transmutable Music control system with structured musical content. The Transmutable song the *Semantic Machine* is presented as a mobile app. The project was developed in collaboration with Florian Thalmann.[3] It is a song that changes based on the weather, time of day and location, like we all do. You can view the pilot user study in Chapter 4. For details about the mobile app, go to the Companion Website.

I will summarise how the music material was organised within the system. Figure 5.25 provides a schematic of the musical design for the *Semantic Machine* project.

COMPOSITIONAL MODEL AND CONTENT:

- Fifty-three musical stems were created and categorised into four thematic groups: Mix1, Mix2, Mix3, and Mix4.
- These four groups can be understood as different themes.
- Of these stems, there are 11 vocal stems with different lyrics and melodies.
- All the stems are organised into eight-bar audio files and segmented into folders by theme and song section.

Semantic Machine - Musical Design

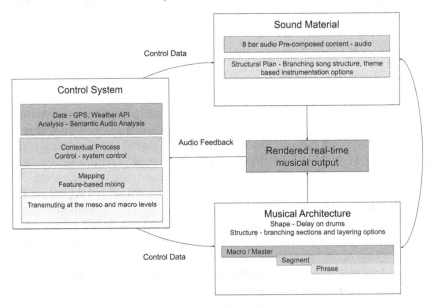

Figure 5.25 Musical Design Model for the *Semantic Machine*.

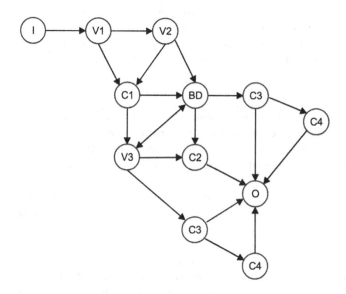

Figure 5.26 Potential Branching Song Structure for *Semantic Machine*.

- They follow a fixed section structure but are adaptable to a branching design for an expandable musical work. See Figure 5.26.
- The semantic player reduces the overall file size by analysing and removing duplicate audio files, making the system more efficient.

MUSICAL ARCHITECTURE

- The music is structured to ensure variability in sections, arrangements, and mixing, with a focus on building energy and intensity.
- A branching structure allows the semantic player to choose the song's progression in real time, influenced by data like the time of day from a mobile phone and the weather.
- Initially composed at different tempos, the works were unified to enable a cohesive blending of tracks.
- A multiscale approach to composition is applied, ensuring that individual stems from all works are compatible and can be layered together in all combinations.

CONTROL SYSTEM

- The semantic player, developed by Florian Thalmann, serves as the control system. It utilises data such as GPS coordinates, weather conditions, location and time to influence musical changes (Thalmann et al., 2016).

- The system autonomously analyses audio files and uses descriptive audio features to determine musical transformations.
- The semantic player's rules and extracted audio features govern how it modifies the track based on the contextual data it receives.

TRANSMUTABILITY OF THE MUSICAL DESIGN

- The semantic player selects which themes to play based on the data recieved, with the ability to combine them in various ways.
- The control system can navigate through different song structures and instrumental combinations, deciding in real time which paths to take based on thedata received.
- The transmutable design allows for varied listening experiences, where the musical output adapts to changes in the user's location, time of day and weather conditions.

Figure 5.27 outlines the structure of the work with four of the mixes. Starting at the master level, the work is organised into song sections, the path of which can be viewed in Figure 5.26. Based on the data received, the system chooses two themes: for example Mix 1 & 2, Mix 1 & 3, or Mix 4 & 4. There are ten different combinations of themes, including the vocals selected. The vocal loops work differently from the instrumental; only one theme is chosen. For example, if Mix 3 and 4 are chosen, a vocal stem from Theme 3 might be played. All vocals also have a randomised control, where the track can be played as an instrumental. The semantic player also selects the number of musical layers based on the time of day. So, the mix will be more ambient at certain times than others.

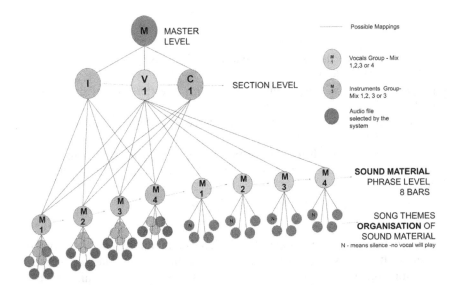

Figure 5.27 Organisation of Sound and Music Material.

Overall, the case study presents a sophisticated approach to algorithmic composition where a control system manages a complex array of musical options, creating a responsive and evolving musical experience for the listener.

Loop Variation (Combining Layering and Branching)

The production of loops has gained popularity in recent years, with loop and sample packs from companies like Splice becoming widely popular. Loops, extensively utilised in electronic music production, form the backbone of many genres that rely on the repetition of loops to build and diminish layers, crafting the music's overall structure. Experienced composers understand that looping an 8-bar or 4-bar drum sequence without any variation is not ideal. Looping with variation is crucial, especially in electronic music, which prevents the listener from growing bored. In game composition, for instance, where a player might be in a level for an extended period, maintaining similar music that loops seamlessly is essential. However, without variation, the player may grow weary of the music.

Interestingly, if the loop varies with each iteration, it can immerse the user more deeply, preventing boredom. Constant change keeps the listener engaged despite the repetition. Consider how you can tune out a recurring sound, like a neighbour's car alarm, once your brain deems it non-threatening; the predictable sound becomes almost imperceptible. Similarly, music with variation keeps the brain engaged and immersed.

Producing loops may seem straightforward, but there are nuances and techniques to streamline the process. For loops to transition seamlessly, they must share the same tempo and conclude in a way that makes their ending indiscernible. In DAWs like Ableton, loops are automatically crossfaded to prevent clicks that can occur if the waveform does not return to zero point at the loop's end, which will be demonstrated shortly.

When creating loops, consider the length of each loop. Do you want all loops to be the same length, such as 8 bars, or do you prefer varying lengths for added variability during playback? The context and medium for which you create the loops should inform whether they need to loop seamlessly, as in a video game environment, or if there should be clear feedback when loops change, alerting the user to the transition. This feedback could come from the music itself or the interface or controls used in the project.

In his book *Writing Interactive Music for Video Games*, Michael Sweet offers valuable insights into composing and editing loops. He describes generating music loops for games as a skill blending musical artistry with technical audio editing expertise. Sweet discusses the concept of clean looping, which is not new; many musicians repeat phrases for a set number of measures, like drum loops. However, software varies in how it supports loop creation, so becoming proficient requires practice to discover best practices in your DAW of choice.

Loops can be seen as musical phrases or sentences. The repetition of these phrases, with variations, can form an entire song section. Your loop could be the length of an entire section or a set of phrases that make up the section, depending on how they are played back and looped together.

Loops are also advantageous for Transmutable Music because they are CPU-efficient. A small amount of music with varied looping can be an efficient composition method, adding randomness and addressing repetition issues. Sweet suggests avoiding dramatic shifts in instrumentation, dynamics, or harmony at the loop point to prevent the loop from being conspicuous to the listener. Instead, musical cadences can make loops more seamless, resolving back to the tonic or root note before the loop restarts.

Loop Editing

When editing loops, there are several factors to consider. Sweet (2015) outlines these as zero-crossing points, waveform shape and direction, transient and legato elements, music tempo, metre, performance, reverb tails, long decays, and crossfading (Sweet, 2015). Let us examine these in more detail.

Zero-Crossing Points: This process is essential, significantly, where audio files are cut at the zero-crossing point. Picture a sine wave: it starts at zero, climbs to a peak amplitude, dips back to zero, drops to a negative amplitude, and then returns to zero. Cutting the audio at zero-crossing ensures that no clicks are heard. If the cut occurs between the amplitude and zero, but the loop begins at zero, a click will be audible. Depending on the sound mix, this can be tricky and may require manipulation and experimentation. Other discussed methods may be necessary if cutting at a zero point is not feasible due to the sound combination.

Waveform Shape and Direction: Imagine the trajectory of a sine wave; if the waveform peaks and returns to zero, the loop should continue this path to prevent clicks. This may require experimentation.

Transient and Legato Elements: Sounds like percussion have sharp transients with a fast attack and quick decay. Others, like synths or strings, have slow attacks and can sustain before decaying. It is generally easier to loop percussive elements due to their transients. For sounds with long decays, fading before the loop restarts is vital. Sweet suggests that percussion at the start of a loop can mask poor looping points, but they should not be over-relied upon. Solutions for challenging loops include placing legato elements away from loop points or fading them out at the end, though this may impact the sound's integrity. Alternatively, overlapping two separate loops so the legato element from one fades into the start of the next can aid seamless looping.

Music Tempo, Metre, and Performance: Looping is more straightforward when loops start and end on downbeats. However, musical styles with swing or offbeat rhythms require careful attention to maintain a seamless loop. Consistency in timing is critical.

Reverb Tails and Long Decays: Many sounds have lengthy reverb tails or decays like legato elements. Consider recording more than you need to find an optimal looping point. Crossfading two loops can allow reverb tails to sound natural at the loop's beginning.

Crossfading: A simple solution for loop issues is crossfading loops together. Whether it involves fading into silence or between two loops, crossfading can create a seamless auditory experience. It is essential to be adept at crossfading two pieces of music smoothly.

Exporting Loops: When bouncing or exporting loops, especially from MIDI to audio, it is prudent to record extra bars to find the best looping point. Always test loops outside the DAW to ensure they do not rely on automatic crossfades that may not be present in the final system.

TIP – Be careful using the MP3 format for loops as it can add a short silence to the start of the loop. It is best to export loops as WAV or AIF files.

Tutorial 5.2 – Composing an 18-loop Work (Phrase Level)

Before beginning this exercise, consider downloading the corresponding Ableton project from the Companion Website.

Your task is to compose three distinct sets of loops; each set contributing to the construction of sections and the overarching structure of the musical output. You should aim to create six loops each for drums/percussion, bass, and a synthesiser or another instrument you choose.

When crafting these loops, infuse them with a rich mix of variation and contrast. This task calls upon your ability to blend layering, variation, contrast, and branching – or, in terms used in game audio, Vertical Orchestration and Horizontal Resequencing.

Reflect on the techniques we have explored in this chapter. Take the six drum loops, for instance: consider employing a range of drum kit sounds or samples for diversity and contrast. Envision how each loop could fit within a song structure – some may be suited for verses, others for choruses. For dynamic range, include variations like half-time beats or 16th-note high-hats. Since you are composing at the phrase level, think about how each loop can contrast and vary, just as a verse does from a chorus.

Your bass and synth parts should be versatile enough to work together in any combination. As you work within Ableton's Session View, remember that creating musical opposites (or contrasts) is essential – call and response, tension and release, light and dark. All these phrases should interplay harmoniously.

Merely producing six MIDI tracks with the same instrument will yield minimal variability. After assembling some ideas in Session View, bring them into Arrangement View. Experiment with layering, subtracting, and combining. Continue refining your ideas, relying on your aesthetic judgement to create numerous possibilities, not just a single option.

What I have found effective in my own process may differ from yours, but I approach composition organically, letting one idea lead to the next. Test your ideas in Arrangement View to see how they unfold over time, ensuring ample contrast and the potential to form new sections. Verify that your loops transition smoothly.

The provided Ableton project showcases the final loops, which are the culmination of several versions. In this project, you will notice the bass loops vary, incorporating synth elements, and even the drum tracks have at least one synth loop. It is not a flawless project, but it should assist you in developing your loop-based track.

Once your MIDI and audio tracks are finalised, consider sampling them down to audio. Exporting numerous loops can be time-consuming, so look for DAW-specific tips. My method involves recording triple the length of bars needed, then looping a portion that transitions well, potentially adding crossfade automation. A handy tip is to start your loop in the middle of the clip to avoid the slow attack of instruments like synths. For an 8-bar loop, record 24 bars.

I have duplicated the Synth loops in the accompanying Ableton set, enabling two to play simultaneously for added variety. Additionally, some return tracks are included. Routing these tracks to the return effects unifies the track and imparts a pad-like texture, facilitating smooth transitions between loops. The loops crafted in this tutorial are designed for use in the mobile tutorial series detailed in Chapter 10.

Summary

Variability is essential for Transmutable Music to allow for unique experiences with each playback.

Variability is achieved through design across all aspects of a Transmutable System, including musical architecture, content, and control systems.

Variation refers to changes within a work's structure or components where some connection to the original is kept.

Variability refers to a work's capacity for change or adaption. It is measured across all components of a Transmutable Music system.

Composing variability requires thinking about multiple options for each part of the music and how they interconnect at all levels.

Variability can be introduced by composing a vast array of materials or focusing on transmutability design.

Simple variability techniques involve changes in music's basic properties and mixing processes.

Complex variability techniques incorporate structured, algorithmic changes and data mapping to control musical elements.

Melodic variation strategies include transposition, reharmonisation, inversion, retrograde, permutation, rhythmic displacement, truncation, expansion, rhythmic alteration, melodic alteration, thinning, and ornamentation.

Layering involves combining distinct melodic, harmonic, and rhythmic elements to create a rich musical texture.

Blending refers to mixing individual sounds to the point where their distinctiveness is lost, resulting in a new, fused sound.

Vertical orchestration is crucial in game audio for modulating intensity and emotion.

Branching and horizontal resequencing showthe structural flexibility required for adaptive music, allowing for real-time composition changes based on in-game events or listener interaction.

A combination of horizontal resequencing and vertical orchestration is often used to create adaptive music.

Variation within loops is essential to keep the music engaging and avoid listener fatigue.

Producing loops requires careful consideration of tempo, ending points, and seamless transitions.

Loops are efficient for Transmutable Music due to low CPU usage and can introduce variation to combat repetitive sound issues.

Loop editing encompasses several technical aspects, such as zero-crossing points, waveform shape, and transient elements.

Crossfading loops can effectively avoid clicks and provide seamless transitions.

The chapter includes discussions, tasks, case studies, and tutorials demonstrating practical approaches to composing variability.

For more practical tutorials, see the advanced tutorials in **Section 4 Compositional Approaches in Transmutable Music: Tutorial Series.**

Notes

1 Mixed and produced by Tim Powles, AKA Timebandt.
2 A radio edit is a version of a recorded song that has been modified or shortened for broadcast.
3 https://florianthalmann.github.io/

References

Erickson, Robert. 1975. *Sound Structure in Music*. Berkeley, CA: University of California Press.
Fat-Boy-Slim. 2004. "Online Mixing Game." BBC Radio 1. 2004. http://www.bbc.co.uk/radio1/fbs/pcexplorer/fbsexppc.html.
Nails, Nine Inch. 2013. "Remix.NIN." 2013. http://remix.nin.com/.

Nelson, Robert Uriel. 1948. *The Technique of Variation: A Study of the Instrumental Variation from Antonio de Cabezón to Max Reger*. Berkeley CA: University of California Press.

Nyman, Michael. 2000. *Experimental Music Cage and Beyond*. 2nd Edition. United Kingdom: The University Press, Cambridge.

Pandora. n.d. "Beyonce – Album – End of Time Stems." http://pandora-news-reviews.co.uk/music/beyonce-knowles/beyonce-songs/end-of-time-stems/.

Sound, Rhythm and. 2003. "Empire Duet." *W/The Artists*. Burial Mix/Finetunes.

SoundCloud. n.d. "Kylie Minogue – Breathe Remix Competition." https://soundcloud.com/groups/kylie-minogue-breathe-remix-competition.

Sweet, M. 2015. *Writing Interactive Music for Video Games: A Composers Guide*. NJ, US: Pearson Education Inc.

Thalmann, Florian, Alfonso Perez-Carrillo, György Fazekas, Geraint A Wiggins, and Mark Sandler. 2016. "The Semantic Music Player: A Smart Mobile Player Based on Ontological Structures and Analytical Feature Metadata." In *Proceedings of the 2nd Web Audio Conference (WAC-2016)*, edited by Alexander Lerch Jason Freeman and Matthew Paradis. Atlanta: Georgia Institute of Technology https://doi.org/ ISBN: 978-0-692-61973-5.

Vella, Richard. 2003. *Sounds in Space, Sounds in Time*. 2nd Edition. London, UK: Boosey & Hawkes Music Publishers Ltd.

6 Transmutability

Transmutability: Reframing Control Systems in Music Composition

Traditional composition practices have always allowed the composer to dictate the qualities of sound, the choice of instruments, the progression of chords, and the intricacies of melody and rhythm. These choices are executed by producers and musicians with their instruments or by composers in a more classical sense. Across history, composers have continually merged new technologies, concepts, and constructs into their music.

Understanding transmutability within a compositional context, however, presents its challenges. Transmutability is fundamentally about change: the extent and capacity for music to evolve or be modified, primarily through the lens of data manipulating musical forms. The relationship between music and data is a familiar innovation, as will be discussed. It lies at the core of numerous music software applications and hardware instruments. When a note is played on a synthesiser, when musical parameters are tweaked, when filters are adjusted, or when a low-frequency oscillator modulates a pitch bend – all these actions are deeply rooted in data, driving the functionality and interactivity at the instrument's heart.

In the context of Transmutable Music systems, 'Transmutability' is a critical metric. It assesses the degree to which input or generated data can alter musical forms.

The term transmutability is referred to in several contexts in relation to using digital data. Scholars such as Lee and Ribas (2016) interpret Transmutability as a creative concept and practice. Their perspective positions it as a transformative mechanism where any data stream becomes a potential source for sonic and visual creation. It is about harnessing data to unfold new interpretations and craft experiences that resonate beyond the traditional senses (Lee and Ribas, 2016).

Hughes and Lang (2006) define Transmutability within the technical sphere as the ability to fluidly modify cultural products encoded in digital formats. The shifting consumption patterns of audiences have given rise to this concept. They separate Transmutability into two main streams: consumer-led and producer-led digital transmutations. These include actions from unbundling

DOI: 10.4324/9781003273554-8

and re-bundling to remixing and sampling, reflecting an era where the boundaries between consumer and producer roles are increasingly blurred (Hughes and Lang, 2006).

Chapter 8 examines the journey from producer-led transmutations to consumer-led transmutations in detail. Changes in technology and communication have created a demand for consumers to engage with production-level transmutations, emphasising the ongoing blending of roles in music creation and consumption.

To broaden this framework, I propose the addition of 'data-led digital transmutation,' which classifies a transformative process controlled by data, capable of altering existing musical forms and generating new ones (Redhead, 2020). This is the embodiment of data as a transformative agent within music.

The design of Transmutability within a music system is multifaceted:

- It requires experimenting with data inputs to explore meaningful musical and sound outputs.
- It calls for investigating how data can embody music within various contexts.
- It necessitates developing a supportive control system to facilitate these explorations.

The design inherently depends on the artist's creative concept and chosen processes. These can range from direct to indirect actions involving human input or autonomous machine interactions.

The data type – discrete or continuous, simple or complex – forms the backbone of Transmutability. It influences everything from the transformation needs to the musical and data analysis within the system, the variability in content and structure, and the extent of data's control within the system.

Control systems vary in complexity. A straightforward design might map basic data transformations onto simple variability parameters such as mixing controls. More intricate designs might incorporate decision-making algorithms, data analysis, and learning mechanisms. These could include generative or AI-driven musical content creation.

Creating Transmutability in a Transmutable Music system extends the variability of the music each time it is played back.

- *Transmutability within the Control System* – The amount and level of control the processes designed within the control system have on the work's form.
- *Transmuting the Musical Architecture* – Data can control the variability designed in the musical architecture and transform and create the structure and form using processes.
- *Transmuting the Content* – Data can control the variability designed in the content and transform and create the content using processes.

In sum, Transmutability is about the capacity of a control system to leverage data in reshaping and creating musical content and architecture. It offers a spectrum of possibilities for composers and listeners to experience and interact with music. Each playback becomes a new discovery through this lens, echoing the system's inherent ability to transform and redefine musical boundaries.

What Comes First, the Data or the Music?

The integration of data into the compositional and production process can be complex. As discussed in Chapter 2, Curtis Roads' methodologies in electronic music composition (top-down, bottom-up, or multi-scale) provide a framework for understanding this complexity.

The top-down approach, while structured, may pose challenges when data is a driving force in shaping the musical architecture. It requires exploratory experimentation to understand how data can yield musically coherent results. Conversely, beginning with a bottom-up approach may lead to an overwhelming array of options, leading to outputs lacking a defined structure or dynamic range. This might suit some composers, but it could be less than ideal for others seeking a more controlled outcome.

A multi-scale approach balances the overarching concept with the granularity of detail and might be more effective – for example, a piece designed to evolve with live weather data. Without first establishing the variability within the composition, predicting the final transmutable effect can be difficult. Similarly, if you employ transmutability as a primary compositional technique, the final sound may remain unknown until the system shaping it is in place.

Ultimately, a musical work will transform into whatever you, as the composer, envision and execute. As you develop the piece, it is common to oscillate between the different approaches (experience, variability, and transmutability), figuring out each part's function at various levels within the hierarchy of the work. This is a process of continual refinement, testing the work within a playback system, and then repeatedly re-evaluating and tweaking each element.

Music as Data and Data as Music

Music can seamlessly be transformed into data. A rudimentary example is transcribing a melody by notating which pitches are played, their duration, and dynamics. While simple melodies might be effortlessly digitised, not all music lends itself so straightforwardly to such a representation.

Conversely, data can be expressed through music. Sonification – the auditory representation of data – is an evolving and captivating field. It offers a unique perspective, letting one 'listen' to data alterations influenced by various factors. This brings to the fore discussions about objectivity and subjectivity in music. It begs the question: is the ensuing music or sound genuinely musical, or is it merely an auditory representation of data? However, even the latter, in its distinct way, might possess aesthetic appeal to a discerning listener.

The rich history of computer music and experimental musical processes provides a sturdy platform to craft ground-breaking musical experiences. At its quantitative core, music is a blend of mathematics and physics, encompassed by frameworks in music theory. The science underlying various types of musical aesthetics is undeniable. Yet, music transcends this as a medium for emotional expression, enabling listeners to feel a spectrum of emotions, from joy to sorrow. Moreover, it expresses identity and communication within societal groups and reflects shifts in art, technology, and society.

In ancient Greece, Pythagoras identified a relationship between the laws of nature and the harmonies found in music (Maurer iv, 1999). The Greek conception of 'music' encompassed a broader spectrum than our modern-day interpretation. To Pythagoras, music and numbers were intrinsically linked. He believed numbers were the underlying key to the universe's physical and spiritual realms. 'The system of musical sounds and rhythms, governed by numbers, mirrored the harmony of the cosmos and corresponded to it' (Burkholder et al., 2019, 843). The musical systems of ancient Greek musicians were rooted in mathematical properties derived from nature. They can be perceived as early formalisms or algorithms, as well as practical applications of numbers or data in contemporary terms.

Musical sounds and rhythms, being numerically ordered, were exemplary of *harmonia* – a concept representing the unity of various parts into a cohesive whole. (Burkholder et al., 2019). This notion, versatile enough to integrate mathematical ratios, philosophical beliefs, societal structure, or specific musical elements such as intervals, scale types, or melodic styles, allowed Greek scholars to view music as mirroring the cosmic order. The connection between music and astronomy was solidified by this concept of *harmonia* (Burkholder et al., 2019).

Ptolemy, distinguished as a foremost astronomer in ancient times, also contributed significantly to music writing. He and others of his time regarded mathematical principles and proportions as foundational to understanding both musical intervals and celestial phenomena. Specific planets, their relative distances, and their orbital paths were believed to correspond with particular musical notes, intervals, and scales (Burkholder et al., 2019). Plato poeticised this concept with his 'music of the spheres' myth, which alludes to the silent symphony created by the orbital dance of the planets. This idea resonated with authors throughout the Middle Ages and beyond, inspiring the likes of Shakespeare in *The Tempest* and Milton in *Paradise Lost*. It also influenced Johannes Kepler, a pivotal figure in modern astronomy, in his work (Burkholder et al., 2019). These ideas came about due to their understanding of number theory and geometry, which makes these concepts relatively simple today. Interestingly, modern radio astronomy is primarily based on detecting radio frequencies in space.

Within this framework, it is apparent how music and data can be understood as interwoven. Although the term 'algorithm' as we know it today did not exist during this era (borrowed from fields like computer science), it is based on a predetermined set of steps devised to address a problem. It also aptly describes the automation process. Such processes for automated composition were employed long before the invention of computers, demonstrating their timeless nature.

Working with data and music offers a multitude of approaches, each with its merit. The choice largely depends on the intended experience for the audience. If the primary focus is a composer's personal journey or scientific exploration, aesthetic considerations might take a backseat. In this context, technology and data can be harnessed innovatively. However, the composer's responsibility remains in crafting a meaningful experience for listeners.

Music Information Retrieval (MIR) has opened avenues to extract various features from a musical piece, transforming it into data. This can range from the emotive content a piece evokes and its tonality to the structural attributes like the onset of sounds or tempo. The nuances can be distinctly identified, such as the register in which instruments play (e.g., a bass guitar's deep tones versus a singer's high pitches). To delve deeper into MIR, see the Moving Forward Resources at the end of this Chapter for links to courses available and tools in audio analysis.

Even if data drives the musical experience, does it remain objective when a composer has crafted a sophisticated system whose features orchestrate the musical narrative?

Exploring Transmutability in Musical Composition

As we navigate the ever-evolving landscape of music and data, control emerges as a multifaceted concept, shaping everything from musical architecture to the systems that govern variability. Such control systems can be likened to complex notations that guide a musical work's transmutability. It is the artist's prerogative to dictate the extent of this control. It can range from intricate melody lines that offer specific variations to allowing broad flexibility without sacrificing quality or artistic vision.

One compelling dimension of control is its delegation. It can be vested in the audience, performers, algorithms, or automated systems. Brian Eno's concept of Generative Music aptly embodies this. He sets the rules and lets the process unfold, embracing unexpected organic qualities and changes in the resulting music (Eno, 1996).

The nature and scope of control can vary depending on the artist's intentions. While one composer might design a system that allows for volume, effects, and mix adjustments, another might strictly curate all possible musical outputs to align with their artistic objectives. However, as explored in *The Madness of Crowds* (Chapter 4), systems may not always operate as envisioned by the artist. This necessitates an ongoing refinement process or, alternatively, a willingness to let the work evolve through user interaction.

This fluidity in control challenges traditional notions of music, especially those held by mainstream artists and audiences accustomed to static, recorded compositions (See Chapter 1). In the world of Transmutable Music, it is not just the end product but the process itself that artists control. This raises an essential question: Can artists adapt to the new skills and paradigms required for composing Transmutable Music systems?

As recorded music begins to shed its static identity, artists and listeners are becoming more receptive to these evolving concepts of control. Far from indicating a relinquishing of artistic authority, the growing flexibility in music making represents a shift in creative and experiential processes.

To summarise, as we explore the potential of Transmutable Music, we realise the need for a new set of skills. This set of skills requires a combination of artistry, technology, and interactive design, which will enable artists to fully realise its transformative power.

Experimentation in Choice and Control

Composers have long been fascinated with offering various degrees of creative control to different individuals involved in the musical process. This could range from simple methods such as dice-rolling to more complex frameworks. (Loy, 2006), points out that many top-tier Hollywood composers delegate tasks to their assistants, thereby relinquishing some degree of control despite having specific criteria (Loy, 2006).

However, music systems[1] developed can include subjective and/or objective choice. Subjective choice can be assigning roles to other composers, performers, or the audience. Objective choice concerns random elements, such as tossing a coin to make musical decisions.

Many contemporary composers incorporate chance elements, leaving room for performers to interpret the score in their own unique ways. This is, of course, a subjective choice as the performer makes creative and aesthetic decisions about the interpretation and representation of the score.

Adding chance and aleatoric options has also been popular for composers. Composers have built compositional systems to explore the results of the system. This can help with compositional ideas or become the work itself. When working with randomness, there are many ways that this can be represented. For instance, the 'urn' and 'random' objects in *Max* 8 allows composers to generate a set of unique random numbers within a specified range, which can be used to add a layer of unpredictability.

The composer's worldview or ontology significantly influences how they embrace or reject these elements of chance in their compositions. I was initially attracted to the concept of Transmutable Music and fluid and amorphous musical forms, perhaps due to my fascination with chaos theory. The idea of relinquishing control and the ripple effects that can emerge from small actions resonated with me deeply. The question then arises: am I employing elements of randomness because I find decision-making challenging, or is it my quest for new discoveries?

Randomness can emanate from various sources, such as natural phenomena. For example, wind chimes are influenced by wind patterns or games decided by dice rolls. Beyond that, some composers believe in the influence of cosmic or natural forces, exemplified by methods such as tarot or the I Ching (Loy, 2006).

Carl Jung coined the term 'synchronicity' to describe events that seem coincidental but are believed to have a meaningful connection (Loy, 2006). Some artists aim to let nature express itself through their work, providing an aesthetic experience even when the message is unclear.

Computers offer various algorithmic methods for generating randomness, from polynomials to linear congruential methods. However, the concept of randomness is only applicable when we cannot define a phenomenon through an exact formula (Loy, 2006).

The field of dynamics studies the impact of forces over time. In a deterministic system, each cause results in a unique effect. However, in a chaotic system, although deterministic, the outcomes appear random. In contrast, a truly random system is non-deterministic (Loy, 2006).

Complexity theory suggests that systems are most stable and adaptable when they are complex. Music, too, benefits from this complexity, engaging the listener's attention. This complexity could form the basis for a new music theory that integrates various disciplines like information theory, chaos theory, complexity theory, cognitive psychology, and nonlinear dynamics (Loy, 2006).

By exploring these multifaceted dimensions, we can better appreciate how the dynamics of chance, control, and complexity influence the ever-evolving landscape of musical composition.

Gareth Joy has authored two books, Musimathics Volume 1 and Volume 2, which offer an in-depth perspective on compositional systems and control that can be utilised for music composition. If you want to gain a deeper understanding of the concepts discussed in this section and enhance your skills, read chapter 9, Composition and Methodology, in Volume 1.

Historical Context

Contrary to popular belief, interactive technologies, algorithmic composition, and musical experimentation are not new phenomena (Collins, 2018; Loy, 2006). These practises have a rich history, stretching back hundreds and even thousands of years. Loy (2006) highlights the Gildo Method, which was developed circa 1026 AD and designed to educate budding composers. This method balanced constraints with the latitude for composers to exercise subjective choices, thereby fostering a framework for pleasing compositions (Loy, 2006). This exploration of systematic music making has evolved into a plethora of music theory tools across various cultural landscapes today (Collins, 2018).

A fascinating contrast arises in the Classical era, particularly the 'Musikalisches Würfelspiel' technique or musical dice games (Loy, 2006). Unlike Guido's method, which centred on human discretion, this technique made chance its cornerstone. The aim was to imbue the composition process with a sense of magic and unpredictability.

In 1757, Johann Philipp Kirnberger published a set of minuets aimed at simplifying the composition process. Although these dice games were promoted as requiring no formal musical training, the initial frameworks were inherently complex, tailored by composers for discerning audiences. Renowned composers of the era, including Mozart, Haydn, and C.P.E. Bach, designed musical frameworks where the throw of a dice could dictate the structure of a piece. The general idea was to present a sequence of musical phrases, the order of which would be determined by dice rolls.

Dietrich Nikolaus Winkel mechanised this concept in creating his 'Winkels Componium,' a barrel organ and orchestrion hybrid. It featured a second barrel that could introduce variations to the primary musical work, providing an extensive range of potential compositions. Loy (2006) points out that this innovation coincided with the era when probability calculus was being pioneered by Pascal (Loy, 2006).

The technology of our time has allowed us to extend these historical experiments in new directions. With modern capabilities, we can design ever-more sophisticated systems for interactive and algorithmic music making. However, it is crucial to remember that the roots of such innovation reach deeply into our shared musical history.

Exploring Data Types for Composition

When I started exploring different data types for this section, I wanted to spark your interest in how data can be used in music composition. However, first, let us clarify what 'data' actually means. Even though we are surrounded by data daily, it is a concept with many layers. Simply put, data is a set of values, which can be numbers or descriptions, that provide some information about things or events.

The word 'data' has many meanings depending on who you ask. For someone in computer science, 'data' often refers to information like bits and bytes that are stored in electronics such as computers and phones. For researchers who study populations, 'data' could mean the numbers collected from surveys like the census. Then there is data science, which is all about making it easier to work with data. It combines math, algorithms, and systems to find patterns in data and uses those findings to make discoveries or decisions.

Data is an umbrella term encompassing numerical values, observations, verbal descriptions, measurements, and a multitude of information forms. It can be separated as qualitative (descriptive) or quantitative (numerical). However, raw data, in isolation, holds minimal significance. Its true power is unlocked when we interpret, model, and derive meaning from it, transforming mere data into insightful knowledge.

Today, data is heralded as a pivotal resource, shaping present and future landscapes. Our personal experiences, as mundane as a social media click or an online search, metamorphose into data, chronicling our preferences and life nuances. Beyond individual actions, there is an avalanche of data from meteorological sources, extensive research fields, and various controllers that can be interfaced and manipulated.

In the field of Transmutable Music, the data spectrum is vast. Some compositions might exclusively utilise data generated from the music itself via analytical systems. In contrast, others might harness environmental metrics like pollution levels, geographic data, or satellite outputs to inform musical transmutability. Other ventures might leverage data generated by users interacting with a specific controller or graphical user interface (GUI).

Recognising that we, as artists and musicians, may not have the expertise of data scientists, this section aims to offer a repertoire of resources and a succinct overview of the methodologies available for engaging with data, deciphering and discovering its intriguing intricacies.

Working with data is multifaceted, and many different methodologies are available at different stages, from collecting the data to interpreting and representing it.

Here is a list of data that could be utilised;

- Surveys and Questionnaires: Standardised tools to collect responses from participants.
- Web Scraping: Extracting data from websites.
- Sensors and Instruments: Collecting environmental, biological, or physical data. This could also include game controllers.
- APIs (Application Programming Interfaces): Tools for fetching data from applications or platforms.

When designing a Transmutable Music system, as illustrated in Figure 3-7, it is essential to recognise that the input data can come from many sources. The examples discussed in this book show that control data can be anything – from car functions, aeroplane controls, GPS coordinates, video game play, weather and pollution data, and even extra-terrestrial data from Mars or the Moon. The type of data you choose will contextualise the musical transformations that occur in your system. Further, who controls this data can significantly impact the experience and the system's adaptability.

Music Representation Systems and Protocols

Music can be represented in a multitude of ways. Its notation has evolved significantly over time. While traditional notation has served for centuries as a conventional medium for capturing and creating compositions, there are other available methods. Since the mid-1980s, the MIDI (Musical Instrument Digital Interface) standard has offered an alternative for digitally representing sound, especially within modern Digital Audio Workstations (DAWs) via the piano roll visualisation feature.

In addition to these more commonly used methods, several specialised languages have emerged to cater to the nuanced demands of musical representation. These include but are not limited to MUSIMAT, CHARM, SCORE, DARMS, GUIDO, DMIX, Kyma, *Max*, PD, Chuck, Supercollider, and Web Audio (Loy, 2006). For those eager to dive deeper into the technical aspects, numerous online courses cover topics from *Max* and PD to JavaScript and Chuck. See Moving Forward Resources at the end of this chapter.

This section will explore simple hands-on tutorials working with MIDI and OSC within the *Max*, *Max4Live* and *Ableton Live* environments. If *Max* and *Ableton Live* are new to you, I strongly recommend exploring some foundational courses (See Companion Website) before progressing to the advanced tutorials in Section 4. Rest assured, the principles and protocols we discuss related to MIDI and OSC are transferable skills applicable across various software platforms and coding languages.

If you are new to working with data and music, I have provided some links in the 'Moving Forward References' at the end of this chapter. However, please feel free to skip ahead if you already know this area.

MIDI Protocol

MIDI (Musical Instrument Digital Interface) is a specification for software and hardware to exchange information. The MIDI protocol was developed in 1983 to standardise all the electronic instruments entering the market. MIDI transformed the electronic music market, making syncing instruments easier.

The protocol is very detailed, but I would like to briefly overview common MIDI messages used to create notes and control parameters. Think of a keyboard. When you press the C note in the middle of the keyboard, you will hear a note with the C pitch. If you press lightly on the key, it will be played softly; if you play it with some force, it will be louder. Also, you can choose how long you would like the note to be played, whether a staccato or short note like a 16th or 8th note. Is it held for longer, like a drone over two bars? MIDI can represent all of these details.

The most common MIDI message sent is 'Note On.' When sending a 'Note On' message, you must specify the pitch, how loud you want to play the note for (velocity), and how long you want to play the note. You must also send a 'Note-Off' message.

Generally, MIDI sends an integer range of 0 to 127. This means there are 128 possible values. So, 128 different pitches and velocities. There are a total of 16 MIDI channels with which to organise your messages.

The 'Change Control' (CC) message is often used to control parameters. Each CC message requires an ID of 0–127 and a value between 0–127. This can be used to control sliders and dials. For example, a fader that controls the gain of a track could be controlled with a change control message – for example, CC ID 1 Value 23. The value will change with continuous data each time the fader is moved up and down. The mobile music tutorial series in Chapters 10 and 11 provides more details on MIDI in practice.

MIDI 2.0

In 2020, the MIDI protocol received a significant upgrade to MIDI 2.0.[2] The potential of MIDI 2.0 is only just starting to be realised. Given the rise in interactive technologies, so many different types of controllers are available to create new expressive instruments and musical expressiveness and experiences. This includes digital instruments being able to map controllers with each other automatically.

I suggest having a look at the new MIDI 2.0 protocol. It offers a vast number of upgrades that go beyond the scope of this book to cover. However, it increases the resolution of controllers, which is something OSC has also achieved. Instead of 128 possible values, there are now 256.

Open Sound Control (OSC)

Open Sound Control is a protocol for network synthesisers, computers, and multimedia devices, like gesture sensors. The OSC protocol is sent over a wireless network. It was developed in 2002 by CNMAT. It can be used as an alternative to MIDI in some cases. http://opensoundcontrol.org/

OSC enables different applications to communicate with each other over a network/ (the internet). OSC can be used to control sound, video, lights, etc.

OSC uses a system like internet addresses. An OSC message comprises an address, type string, and arguments.

- Address – 168.1.1.0 (IP type)
- Type string

  - i integer (int)
  - f float
  - s string
  - b blob

- Arguments – zero or more

Example - 168.1.1.0/oscillator/2/frequency, f 440

The address is a tree using slashes '/'. The comma means it is an argument. The 'f' means it is a float argument, of 440.

OSC is a content format. It uses UDP User Datagram Protocol. OSC travels over the UDP network. Other protocols used on the internet are FTTP, TCPIP, and HTTP. There are lots of different ways to transfer data over the internet.

The sender requires an IP address and a port number. The receiver only requires a port number. The receiver is already at an IP address. The receiver port number is exclusive. Think of a port as the number of a storage shed where your things are stored. The port number is where your information is stored within your computer. You will not find your data if you listen to the wrong port. It is important to note that no other receiver can use it once a receiver starts listening to a port. Tutorial 10.1 provides more details on OSC in practice.

Tutorial 6.1 Chance and MIDI Tutorial: Utilising Randomness in *Max4Live*

In this tutorial, we will create a randomised melody in C major using *M4L* and *Ableton Live*.

M4L provides several objects for random probability: 'random,' 'choice,' 'urn,' and 'drunk.' Please look at the help patch for each object for more information on how to use them. You can access a completed version of this tutorial on the Companion Website. The *M4L* object and *Ableton Live* set can be downloaded.

1 Start by opening a new *Ableton Live* set.
2 Drag a blank *M4L* MIDI effect into your first MIDI track. See Figure 6.1.
3 Open your new blank *M4L* **MIDI device** in *Max* using the third button from the left among the right-hand side buttons on the device.

'makenote' and 'noteout' objects

4 Create a new object by pressing 'n' and typing 'makenote' to generate a 'makenote' object.
5 This 'makenote' object accepts three types of messages: the first is pitch, the second is velocity, and the third is duration. We will focus on pitch and velocity to produce sound in *Ableton Live*.
6 Press 'm' to create a new message box and type '60,' corresponding to middle C in MIDI. Connect the output of the message box to the left inlet of the 'makenote' object, as per Figure 6.2.
7 Create another message box with 'm' and type '100' to set the velocity, dictating the note's loudness. Connect the output of the message box to the middle inlet of the 'makenote' object, as per Figure 6.2.
8 Next, press 'b' to create a bang object. Connect the bang to the left inlets of the two message boxes, like in Figure 6.2.
9 To send the MIDI note, we need a 'noteout' object. Press 'n' to create a new object and type 'noteout.' Now patch the left outputs of the 'makenote' object to the left inlet of the 'noteout' object, and the right outlet of the 'makenote' to the middle inlet of the 'noteout' object, as shown in Figure 6.2.

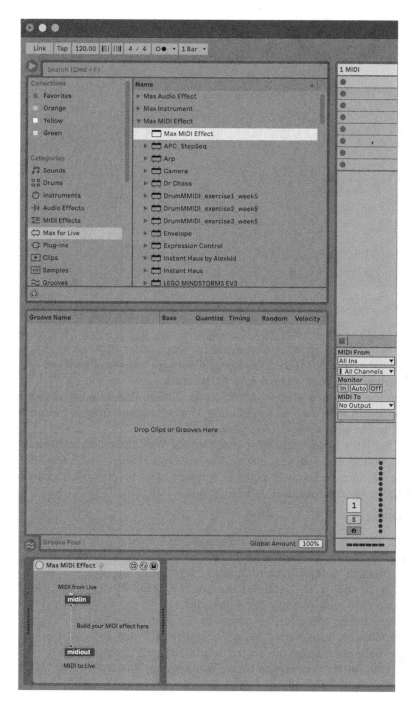

Figure 6.1 Screenshot of *Ableton Live*.

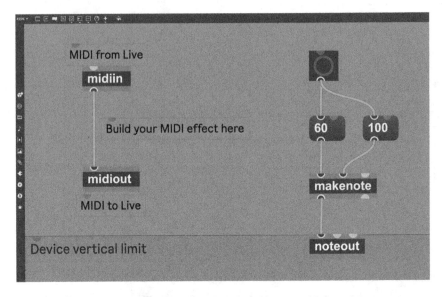

Figure 6.2 Screenshot *M4L* Patch.

10 Lock the patch by pressing Command (or Control for PC) and 'e.' Click on the bang to send MIDI data. **You will not hear any sound until you add an instrument to your *Ableton Live* MIDI track.**

11 In *Ableton Live*, use the browser to find the 'Grand Piano Pad.adg' under Instrument Rack/Piano & Keys/Grand Piano Pad, and drag it onto the MIDI track, **positioning it after your** *M4L* device.

12 Click the bang within your *M4L* patch again. With the piano instrument added, you should now hear a note. The 'makenote' object will handle the timing of notes, sending a note-off message after a specified duration. Which, at the moment, is a default setting.

'metro' Object

13 Ensure you are back in edit mode by pressing Command (or Control for PC) 'e.'

14 Now, create a method to continuously trigger notes by adding a 'metro' object, type 'metro 500.' Adding the argument 500 sends a bang every half a second (500 milliseconds). See Figure 6.3.

15 Patch the outlet of the 'metro' object to the inlet of the 'bang.'

16 A shortcut to create a new 'toggle' object is 't.' Patch the outlet of the 'toggle' into the inlet of the 'metro' object, as per Figure 6.3.

17 Lock your patch, turn on the toggle and test that you can hear a MIDI note every half a second.

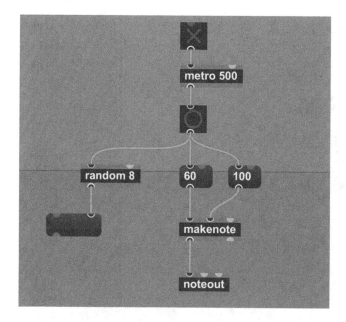

Figure 6.3 Screenshot *M4L* Patch.

'random' object

18 To introduce pitch variety, create a 'random 8' object; see Figure 6.3. Adding the argument '8' will create a random number ranging from 0 to 7. This will create a randomised number generation for the eight notes in the C major scale (within the range of one octave).

19 Connect the outlet of the 'random' object into the right inlet of a new 'message' box ('m'); see Figure 6.3.

20 Next, patch the outlet of the 'bang' object into the left inlet of the 'random' object, see Figure 6.3.

21 Lock your patch and turn on the 'toggle.' You will hear the MIDI, but check the message box and observe that a random number between 0 and 7 is selected each time. Turn off your 'toggle' and unlock your patch.

'select' object

22 With the 'select' object, we can access each number separately. We want to list the MIDI notes representing our 0–7 scale degree number. Create a new 'select' object. We need to tell the select object which numbers we want it to pay attention to. Add the numbers that are being randomised 'select 0 1 2 3 4 5 6 7,' as per Figure 6.4.

23 The 'select' object will create eight new outlets. For example, if 0 is randomly selected, we can bang a specific number. Create eight message

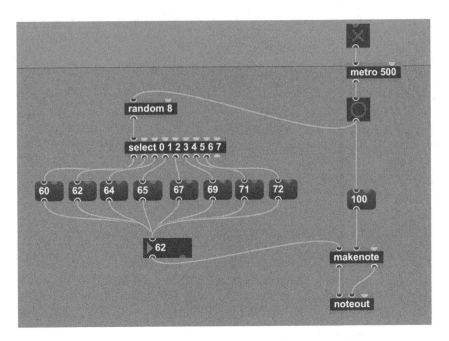

Figure 6.4 Screenshot *M4L* Patch.

objects by pressing 'm.' In each one, we will type in the number of the MIDI note we want to send. Here, we can use a little bit of beginner's music theory.

A major scale comprises of notes that are intervals of a tone (T) or semi-tone (S). (A tone is, of course, two semitones.) The formula for the major scale is TTSTTTS.

If we start at note 60, the following note in the scale would be a tone away (or two semitones), so the following number in the sequence would be 62. The following note is also a T away and would be 64; the fourth scale tone would be 65 as it is only a semi-tone away. Create an integer object by pressing 'i.' In each message, write the notes you want to send (60,62,64,65,67,69,71,72). Join the objects together, as shown in Figure 6.4. Go out of edit mode and press the toggle button. You can see the number sent each time in the integer object.

24 To hear the notes, delete the link to the original '60' message and replace it with the outlet of the integer, as shown in Figure 6.4.

'drunk' object

25 Let us now add some variation of the velocity to add dynamics and a more human-like quality. As with everything in *Max* and *M4L*, there are so many ways to do the same thing. We will use the 'drunk' object. Create

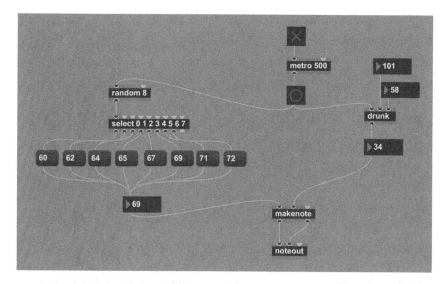

Figure 6.5 Screenshot *M4L* Patch.

a new object and type 'drunk.' If you want to learn more about using the 'drunk' object right, right-click, and press help.

26 For the 'drunk' object to work, we must provide a number range and the number of steps the following random number selected could take. To do this, create three integer objects by pressing 'i.'

27 Patch the first integer box outlet to the middle inlet of the 'drunk' object. The middle inlet is the range.

28 Patch the second integer object outlet to the right inlet of the drunk object. This is the number of steps. See Figure 6.5.

29 Next, as per Figure 6.5, patch the outlet of the 'drunk' into the inlet of the third integer box. Now, we can see what numbers are being sent out.

30 Finally, connect the outlet of the bang to the left inlet of the 'drunk' object AND the outlet of the bottom integer box into the middle inlet of the 'makenote' object; see Figure 6.5.

31 Exit edit mode and adjust the values in the integer objects, instructing the drunk object. Remember, the velocity range goes up to 127. Use 101 for the range. Add 58 to the number of steps. Experiment with different values, 8 in the steps, for example.

Note Duration

32 We will now create variations of the note length or duration. Have a go at doing this yourself before checking the instructions. Hint: use the 'random' object.

33 Create a new object and type 'random 3' – see Figure 6.6.

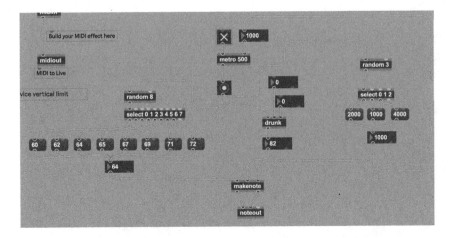

Figure 6.6 Screenshot *M4L* Patch.

34 Next, create a new 'select' object with the arguments 0 1 2. See Figure 6.6.
35 Patch the outlet of the 'random' object' to the left inlet of the 'select' object.
36 Create three new messages, 'm,' and add the values 2000, 1000, and 4000. You can change these values for different instruments. It depends if you want a fast or slow melody.
37 Patch the three most left outlets of the 'select' object outlet from the left of the select object into the left inlet of the three message boxes.
38 Now, patch the message outputs into a new integer 'i' object.
39 Next, connect the outlet of this integer box into the right inlet of the 'makenote' object. This will set the note duration in milliseconds. See Figure 6.6.
40 Connect the outlet of the bang into the inlet of the 'random' object.
41 Finally, we will need to connect the outlet of the integer box into the right inlet of the 'metro' object. Alternatively, you could connect it to the input of the integer box connected to the 'metro' object to check the values. This step means that the metro does not send a bang every 500 ms; it will send a note at the end of each note duration.

Presentation Mode

We will now get our *M4L* device looking better. We can do this with presentation mode.

42 Right-click on the toggle object - make sure you are in editing mode. Select 'Add to Presentation.' See Figure 6.7.
43 You will see a screen icon at the bottom menu of your patch. Click this so it is yellow. This will take your patch into presentation mode. Adjust your

Figure 6.7 Screenshot *M4L* Patch.

toggle into the centre on the left of the patch. So you can see it when the
device is open in *Ableton Live*.

44 Now, we need to tell the patch to open in presentation mode. When in
editing mode, right-mouse click anywhere in the patch (not on an object)
and select 'Inspector Window.' In the Inspector Window (view selection),
tick the 'Open in Presentation' box as per *Figure 6.8*.

45 Now save your *M4L* device; you will see the toggle in your device view.
To make the MIDI notes work, you need to toggle the device on. Also,
ensure there is a MIDI instrument after the device in *Ableton Live*. See
Figure 6.9 to see the finished *M4L* device.

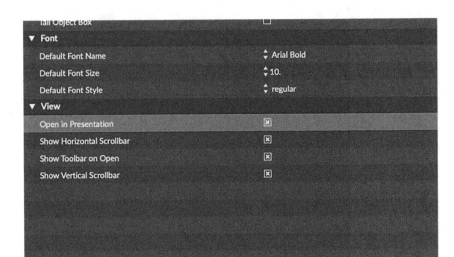

Figure 6.8 Screenshot *M4L* Patch.

Figure 6.9 Screenshot *M4L* Device.

Experimenting and Tweaking

46 Next, duplicate the MIDI track with the *M4L* device we have created and change the instrument to a more ambient sound.

47 You could increase the note durations, so you get more drone-sounding tones.

48 Try adding a bass instrument. Copy and paste your *M4L* Device to other tracks and 'save as.' With the Bass, try reducing the note pitch by two octaves. Add an object and type '- 24' remember that an Octave is 12 semitones.

49 Finally, add a drum kit track and experiment with trying to adapt the note duration and triggering pitch to match the kick in the rack. It is usually MIDI number 35 – C0, but some kits are designed differently.
50 You can download the completed *Ableton Live* set in Tutorial 6.1 from the Companion Website.

Data Transformation, Modelling, and Mapping

In the second chapter, the types of control were categorised as follows:

- **User Control:** Audience interaction through a controller or graphical user interface (GUI).
- **Machine Control:** Musical analysis and algorithms, including machine learning, deep learning, and semantic technologies.
- **Chance Control:** Utilisation of chance operations and randomness.
- **Contextual Control:** Use of research or predefined data lists, internet data via APIs (e.g., Google Maps, WeatherAPI), or sensor-based information such as gestures, accelerometers, gyroscopes, and conductivity measurements.

The second chapter established a direct relationship between the type of control and the music system employed, such as interactive, reactive, adaptive, generative, autonomous, responsive, contextual, algorithmic, or AI-driven systems, noting that combinations are common. For instance, the *Semantic Machine* (see Chapters 4 and 5) project merges contextual, AI, and autonomous elements using machine and contextual controls.

The approach to transforming and mapping data into musical elements varies with the data's complexity. For example, datasets with a single value are relatively straightforward to represent. This is illustrated in the tutorial in Tutorial 10.2, where we explore mapping basic data inputs from a mobile phone's *TouchOSC* interface, such as button presses and slider movements, onto control aspects of a musical piece.

Working with complex datasets often requires advanced techniques to map them effectively. The value we get from data comes from how we represent and interpret it. Take, for instance, the intricate sensor data from devices like the R-IoT from IRCAM. The data captures and sends detailed motion data such as movements and orientation through OSC. This type of data can be pretty challenging to understand. I remember spending months during my PhD trying to figure it out. It was a challenging but enlightening period where I tried to create a tool in *Max4Live* to translate dance movements into music. My initial attempt was not great, mainly because I was still learning, and the task was complex, involving spatial data and vector math – things like quaternion angles, which are tricky unless you are familiar with that kind of math.

Thankfully, machine learning has become a game-changer for people like me. It offers tools that can learn from the data we give them, which is a huge help for artists. Tutorial 11.3 in Chapter 11 covers the basics of creating models with machine learning for different data types. Furthermore, if you want to go even deeper into this topic by using various software and sensors, as well as learn about capturing dynamic movements over time, I recommend Rebecca Fiebrink's course on Machine Learning for Musicians and the *Max* patches from the Fluid Corpus Manipulation (FluCoMa) project. You can find links to these resources in Moving Forward Resources later in this chapter.

Mapping to Musical Parameters (Simple)

Music intrinsically includes numerical aspects: it expresses notes, rhythms, and tempo in numeric terms. Thus, it is a logical step to use numerical data from external sources, whether mathematically generated or observed from phenomena, to influence musical elements. The mapping process translates this data into music, a central technique in algorithmic composition and data sonification (Ransbeeck, 2015).

Whether predetermined, algorithmically produced, or derived from a model, data requires mapping strategies to translate it into musical parameters. This mapping is a crucial operation within Transmutable Music's control systems.

The challenge lies in determining how data will influence the composition. It is about interpreting the data in a way that meaningfully contributes to the piece. Will it dictate the pitches in a melodic phrase? The rhythm of a drum pattern? Deciding on how data is used in a work is essential for effectively integrating it into a musical context.

For instance, Tutorial 6.1 demonstrates mapping randomised controls to a scale system that can produce notes of varying lengths and velocities. Such control could alternatively be initiated through a GUI when a user interacts with an app or be responsive to datasets like brain activity or Martian wind speeds.

Mapping data to musical parameters can be done in various straightforward ways:

- **Direct Mapping (One-to-One):** For example, moving a slider on a MIDI controller can directly adjust a single parameter, like the mix between wet and dry signals of a reverb effect in a Digital Audio Workstation (DAW).
- **Direct Mapping (One-to-Many):** A single data input, like a MIDI controller slider, can simultaneously control multiple parameters in a DAW, such as the reverb mix, the frequency of an auto-filter, and the volume of an instrument.
- **Direct Mapping (Many-to-One):** In interactive scenarios, multiple inputs, like pressing two distinct buttons by audience members, can combine to

trigger a single response, such as starting a loop (Wanderley and Depalle, 2004; Drummond, 2009).

When the mapping is not straightforward due to the complexity of the data, it is important to consider data transformation and scaling. For instance, with continuous data that needs to be mapped to a DAW control dial, you must match the data's range to the dial's requirements. If the dial operates on MIDI values (0–127), the data should be converted to integer values within this range. Alternatively, if the dial works on a scale from 0 to 1, the data should be formatted as a floating-point number (like 0.3983), allowing for more nuanced control with a higher resolution.

It is important to note that not all mappings that represent one-to-one mappings are simple. Carla Scaletti highlights that directly converting data into sound can be as complex as it is insightful, using an example of tidal data. She explains that if you collected data on ocean tides every hour for eleven years, you would have a vast amount of information, similar to how a digital audio track is made of many sound snapshots (Scaletti, 2018).

In digital music, these snapshots, known as the sample rate, are taken 48,000 times every second. If you were to convert your tide data into sample rate values and play it, you would whizz through eleven years of data in a mere two seconds. Seasonal tides would produce slow hums, while daily tides would sound like high-pitched notes (Scaletti, 2018).

However, if your data is incomplete, irregular, or imprecise, the sound could end up distorted or noisy. However, when converted accurately, listening to data turned into sound can reveal patterns and changes that are hard to see but easy to hear (Scaletti, 2018).

Scaletti suggests not rushing through the data but instead slowing it down, playing only forty-eight points every second. This would stretch the cycles out. The daily tides would happen a few times every second, and seasonal changes would take minutes to complete. While these slow changes are too gradual to be heard as pitches, they can be used to alter the sound of an audio synthesiser in meaningful ways (Scaletti, 2018).

For instance, you could use the tide data to adjust the tone of a synthesiser. Imagine the tide data acting like a hand on a knob, turning it slowly up and down. With each rise and fall of the tides, the sound would change, maybe becoming fuller or thinner. This is a different kind of data-to-sound conversion, where the data is not the sound itself but affects the sound, allowing you to 'hear' the tides by listening to the changes they make in the synthesiser's noise. It is a clever way to make the patterns in the data audible and understandable, even if they are happening too slowly to hear directly (Scaletti, 2018).

Despite these challenges, when you manage to convert data to sound correctly, it can tell you a lot. Our ears are good at picking up on rhythms and patterns, even without any special skills. This means we can often hear things in the data that we might not see just by looking at numbers or graphs (Scaletti, 2018).

As described by Scaletti, the processes involved in working out the best way to sonify data are experimental and iterative. Beyond mapping data directly to musical parameters, we can use modelling.

Data Modelling and Mapping: An Overview (Complex)

When discussing incorporating data into music, we look at ways to translate numbers and measurements into sounds and musical expressions. Different methods exist, depending on whether the data changes over time or stays constant.

Doornbusch discusses mapping in Algorithmic composition. He provides examples of two important historical works. In the composition, *Pithoprakta* by Iannis Xenakis, the movement of gas particles modelled through Brownian motion and the Law of Large Numbers informs the structure of the music. Xenakis graphed over a thousand calculated particle velocities at given times on an XY plane, translating them into glissandi for 46 string instruments, mapped across 15 major third intervals corresponding to their pitch ranges. The mapping is direct, with uniform intensities and durations, but temporal complexity creates the sensation of a particle cloud. This approach reflects Xenakis' broader practice of using direct mapping from datasets to musical parameters (Doornbusch, 2004).

Charles Dodge's *Earth's Magnetic Field* also exemplifies algorithmic composition through data mapping. Using a Bartels diagram, Dodge translated the variations in Earth's magnetic field, measured by the Kp index, into pitches and rhythms over a compressed timeframe. While pitches directly correlate with the Kp index levels, timbres were aesthetically chosen, and spatialisation speeds, directions, and rhythms were determined by plotting data section lengths against *Max*imum amplitudes. The data's multiple readings and resemblance to 1/f noise contribute to the composition's self-similarity (Doornbusch, 2004).

These cases illustrate varied mapping approaches in algorithmic composition, where data can be a direct, linear influence or part of a more intricate nonlinear process, achieving structural cohesion within the musical work (Doornbusch, 2004).

As discussed in Chapter 2 in more detail, there are distinct data forms for discrete (one-off event), continuous (constant data stream), and hybrid data types. Many of these models will be familiar to those familiar with algorithmic composition. I will discuss a variety of models that are available to work with different data types.

Models do not need to be complicated. For example, using a decision stump or tree is a logical start. An example of a decision stump is if the data equals or is greater than 0, create an event, or send a bang.

Discrete Data Modelling

Discrete data models often involve creating events that correspond to specific data points. For instance, one might programme an event to occur when the temperature exceeds 35 degrees Celsius, such as triggering a sound clip or engaging an effect that evokes a summery vibe.

Standard models for discrete data include:

- **Markov Chains**: These predict the likelihood of a sequence of events based on past data.
- **Decision Trees**: These models use a structure of if-else statements to determine actions based on data values. For example, a value that meets a particular condition could trigger a specific sound or effect. In Tutorial 11.2, we explore using decision trees to generate events from accelerometer data. Decision stumps only involve one if-else statement.

Constructing models for data interpretation can be simple. Starting with something as straightforward as a decision stump or tree can be effective. For instance, a decision stump could be programmed with a simple rule. If the incoming data is zero or positive, trigger an event or initiate a signal, known in the musical programming language *Max* as a 'bang.'

Machine learning algorithms can offer more nuanced user experiences than basic decision tree approaches, as they can efficiently handle complex decision-making processes. Nonetheless, decision trees are suitable for straightforward data modelling tasks.

Dealing with Noise in Data

When working with continuous data, filtering out noise – irrelevant or redundant data – is crucial. Machine learning models can be particularly effective in identifying and mitigating noise. Alternatively, one can manually curate the dataset to remove any extraneous information.

Continuous Data Modelling

Modelling continuous data presents its own set of challenges, mainly due to the rapid changes in values that can occur. A fundamental approach is to scale the data and map it to a control element like a slider or a dial, which is more suited to continuous input than a button. Some strategies for modelling continuous data include:

- **Creating Classes or Events** involves segmenting continuous data into distinct categories or triggers.
- **Regression**: Often used in machine learning, regression deals with continuous data, identifying trends and predicting values.

It is important to note that machine learning is not limited to either discrete or continuous data. It can handle both types, as well as time-series data. Different algorithms are designed to work best with specific types of data and desired outputs. For instance, will the output be discrete classes, or will it involve continuous data streams?

Machine Learning: Simplifying Complex Data

Machine learning simplifies interpreting data by training a model to represent your inputs accurately. It is versatile and can work with various data types, whether dealing with discrete categories, continuous fluctuations, or temporal patterns. The choice of algorithm will depend on the nature of your data and the kind of output you aim to generate, whether that be discrete classes or continuous variables.

Applying these concepts in practice can demystify the seemingly complex data modelling process and harness it effectively for creative purposes.

Understanding these concepts and their application to music enables us to explore all the creative possibilities that data offers. Whether adding a new dimension to your live performances or crafting an interactive installation, these tools open up a new frontier in musical expression.

This book primarily addresses basic data transformation techniques. For those interested in more complex applications, Data Sonification and Algorithmic Composition fields provide an extensive range of intriguing works and frameworks.

For hands-on application of the theoretical concepts covered, refer to the tutorial series in Chapter 11. This series provides detailed guidance on fundamental and advanced data mapping techniques, introduces machine learning principles, and demonstrates how to utilise your phone's accelerometer sensors for musical purposes. Further exploration on creating and integrating adaptive music within gaming environments, including how game-generated data can influence music playback, is available in Chapter 12.

If you are ready to dive into practical experimentation, I have developed a series of *Max* patches tailored to various data types and methodological approaches. These patches are designed to jumpstart your projects and can be easily downloaded through the Companion Website.

1 Accessing a data set in *Max*
2 Accessing data from the internet via a weather API in *Max*
3 Markov chain for drum trigger device in *Max4Live*

AI and Creative Data Utilisation

Artificial Intelligence is rapidly beginning to transform our world. It offers the capacity to interact with data in new and innovative ways. AI can detect patterns, nuances, and relationships within data that would be either difficult or unobvious for humans.

- AI can generate new content, whether it is a piece of music, artwork, or a written piece, by learning from vast datasets.
- AI can convert datasets into auditory experiences, allowing for a different perspective on data interpretation.
- By analysing multiple music tracks, AI can craft a new piece by blending elements from them, thereby reshaping the auditory landscape.
- AI can craft music or visuals tailored to an individual's preferences, moods, or even biometric data.
- With real-time data processing, AI can alter music or visuals based on user interactions or environmental factors.

It is predicted that AI's potential lies not just in processing data but in its capability to intertwine creativity into it, enabling novel forms of artistic expression and experiences. The debate around creativity and AI is multifaceted and not one I mean to poke. In my personal experience, the complexity of human emotion, problem-solving, and perception is one of the essential ingredients that makes art and human creativity so authentic.

As technology, especially AI, continues to evolve, our ability to process, interpret, and act upon vast amounts of data grows. However, this commodification of personal data is concerning. With every click, every interaction, and every online transaction, individuals generate data footprints. As this data accumulates, its potential value – both economically and strategically – increases. This scenario is reminiscent of the gold rushes of the past, but with personal data being the sought-after resource.

The challenge is balancing the benefits of these advancements with ethical considerations and privacy protections. Just as the gold rush brought opportunity and exploitation, the 'data rush' presents similar challenges. Ensuring that AI serves as a tool for the betterment of society, rather than just a means of profiteering, is essential as we navigate the next digital frontier.

Historically, artists have been at the forefront of societal reflection, offering critical perspectives on emerging trends, paradigms, technologies and challenges. Integrating data within a compositional approach offers new avenues for creative expression and serves as a commentary on our data-driven existence.

As artists experiment with data in their compositions, they can illuminate the ubiquity of data collection, highlight its implications, and challenge prevailing norms about data privacy. This form of artistic expression becomes an intersection of creativity and advocacy, urging awareness and introspection.

Moreover, artists can make abstract concepts tangible by incorporating data into their art. They can sonify or visualise the otherwise invisible and intangible data flows, making the abstract personal and experiential for their audience. In doing so, they bridge the gap between the esoteric world of big data and the everyday experiences of individuals.

Moving Forward Resources

Here, I have provided a short selection of resources that could help in several areas within Transmutable Music. Many links are courses on the Kadenze platform. It is a fantastic training platform for anyone at any level. I have also provided essential links for programming and software. Check the Companion Website for more links.

Learning Audio and Programming Software

Max MSP Programming Course: Structuring Interactive Software for Digital Arts
https://www.kadenze.com/courses/Max-msp-programming-course-structuring-interactive-software-for-digital-arts/info

Ableton Live Course: Sound Production for Musicians and Artists
https://www.kadenze.com/courses/sound-production-in-AbletonLive-live-for-musicians-and-artists/info

Learn *RNBO*
https://rnbo.cycling74.com/learn

Learning *M4L*
https://help.AbletonLive.com/hc/en-us/articles/360003276080-Max-for-Live-learning-resources
https://docs.cycling74.com/Max8/vignettes/live_resources_tutorials

Creating Interactive Audio Applications Using *Pure Data*
https://www.kadenze.com/courses/creating-interactive-audio-applications-using-pure-data-i-i/info

Introduction to Real-Time Audio Programming in *ChucK*
https://www.kadenze.com/courses/introduction-to-programming-for-musicians-and-digital-artists/info

Real-Time Audio Signal Processing in *Faust*
https://www.kadenze.com/courses/real-time-audio-signal-processing-in-faust/info

Web Coding Fundamentals: HTML, CSS and Javascript
https://www.kadenze.com/courses/web-coding-fundamentals-for-artists/info

Loop: Repetition and Variation in Music
https://www.kadenze.com/courses/loop-repetition-and-variation-in-music/info

Analysis, Learning Algorithms, and Model-Based Approaches

Introduction to Sound and Acoustic Sketching
https://www.kadenze.com/courses/introduction-to-sound-and-acoustic-sketching/info

Generative Art and Computational Creativity
https://www.kadenze.com/courses/generative-art-and-computational-creativity/info

Physics-Based Sound Synthesis for Games and Interactive Systems
https://www.kadenze.com/courses/physics-based-sound-synthesis-for-games-and-interactive-systems/info

Machine Learning for Musicians and Artists
https://www.kadenze.com/courses/machine-learning-for-musicians-and-artists/info

Extracting Information from Music Signals
https://www.kadenze.com/courses/extracting-information-from-music-signals/info

Machine Learning for Music Information Retrieval
https://www.kadenze.com/courses/machine-learning-for-music-information-retrieval/info

Fluid Corpus Manipulation – *Max* External, machine learning, audio analysis
https://learn.flucoma.org/

Summary

Transmutability challenges traditional music composition by incorporating change through data manipulation.

Music has long been intertwined with data, from ancient Greek theories linking harmonics and astronomy to modern digital synthesis.

Different data types, such as environmental readings or user interactions, offer new dimensions for composing music. Data is a broad term that includes various forms of information. It can be qualitative or quantitative and requires interpretation to unlock its potential.

Music notation has evolved, with systems like MIDI and OSC protocols enabling complex digital representations and interactions.

The concept of control in music varies, from systems that strictly guide the musical output to those that incorporate randomness and chance.

Max4Live includes objects like Random, Choice, Urn, and Drunk to introduce probabilistic elements into music composition.

Mapping is crucial for translating data into music.

Data modelling can be for discrete or continuous data, and machine learning can aid in simplifying complex datasets. Different models and algorithms are available to handle various data forms for musical composition.

AI's role in music is expanding, and it is capable of generating new compositions, sonifying data, and creating personalised experiences.

The chapter encourages considering the ethical implications of data use, privacy, and AI in the broader context of creativity.

The historical context shows that interactive and systematic music making has roots that predate modern technology.

Notes

1 A music system, in relation to choice and control, is an arrangement that allows performers or listeners to influence the direction and elements of the music through interactive decisions and manipulations.
2 https://www.midi.org/midi-articles/what-musicians-and-artists-need-to-know-about-midi-2-0 or for the whole manual, https://midi.org/specifications/midi-2-0-specifications/midi2-core/midi-2-0-core-specification-collection

References

Burkholder, J. Peter (James Peter), Donald Jay Grout, and Claude V. Palisca. 2019. *A History of Western Music. Tenth International Student Edition.* 10th Edition. New York: Norton and Company.
Collins, Nick. 2018. "Origins of Algorithmic Thinking in Music." In *The Oxford Handbook of Algorithmic Music,* edited by Dean, R. T. McLean, A, 67–78. New York: Oxford University Press.
Doornbusch, Paul. 2004. "A Brief Survey of Mapping in Algorithmic Composition." In Proceedings of the International Computer Music Conference (2002), http://www.academia.edu/download/33447946/A_Brief_Survey_of_Mapping_in_Algorithmic_Composition.pdf
Drummond, J. 2009. "Understanding Interactive Systems." *Organised Sound* 14 (2): 124–133. http://search.proquest.com.ezproxy.newcastle.edu.au/docview/1554438?accountid=10499.
Eno, B. 1996. "Generative Music." http://www.inmotionmagazine.com/eno1.html.
Hughes, J, and Karl R Lang. 2006. "Open Source Culture and Digital Remix: A Theoretical Framework." *Journal of Management Information Systems* 23 (2).
Lee, Catarina, and Luísa Ribas. 2016. "Exploring Textual Data: Transmutability as a Creative Concept and Practice." In *Proceedings of XCoAx 2016 Fourth Conference on Computation, Communication, Aesthetics and X.* Bergamo, Italy: Communication, Aesthetics and X. http://2016.xcoax.org/pdf/xcoax2016-Lee.pdf.
Loy, Gareth. 2006. *Musimathics, Volume 1: The Mathematical Foundations of Music.* Cambridge: The MIT Press.
Maurer iv, John A. 1999. "The History of Algorithmic Composition." CCRMA Stanford University. 1999. https://ccrma.stanford.edu/~blackrse/algorithm.html.

Ransbeeck, Samuel Van. 2015. "Composition with Complex Data: A Contribution on the Mapping Problem Through Practice-Based Research." *PQDT - Global*. Portugal. https://www.proquest.com/dissertations-theses/composition-with-complex-data-contribution-on/docview/2033065632/se-2?accountid=14681.

Redhead, Tracy. 2020. *Dynamic Music: The Implications of Interactive Technologies on Popular Music Making*. NSW, Australia: University of Newcastle. http://hdl.handle.net/1959.13/1413444.

Scaletti, Carla. 2018. "Sonification ≠ Music." In *The Oxford Handbook of Algorithmic Music, Oxford Handbooks*, edited by Roger T. Dean and Alex McLean, 263–385. Oxford: Oxford Academic.

Wanderley, Marcelo M., and Philippe Depalle. 2004. "Gestural Control of Sound Synthesis." *Proceedings of the IEEE* 92 (4): 632–644. https://ieeexplore.ieee.org/document/1278687.

7 Evaluating Transmutable Music

Practitioner Evaluation and Roles

Assessing the value of a music work can be a complex endeavour, tangled in the webs of personal taste and perception. However, we can rigorously analyse a piece's process, creation, and experiential elements. When it comes to music leveraging interactive technologies, the complexity intensifies, demanding a tailored model for evaluation. Despite the growing interest in this field, I contend that the full potential of transmutable music remains untapped. This is primarily due to the intricate development, composition, and production process, all of which require unique playback systems for proper appreciation.

Researcher evaluation has been instrumental in supporting the emerging domains of interactive art and music interaction, focusing on user experience and system functionality to refine the work and enhance overall engagement (Candy and Ferguson, 2014; Holland et al., 2013). This approach is increasingly relevant, even though not all Transmutable Music endeavours necessitate active audience participation.

Typically, the music itself escapes judgement. Instead, the focus is on evaluating the system, the user experience, and the user's perception. This chapter introduces a framework aimed at assessing the musical design alongside the supportive capacity of the system. It can be used to support the further development of software, tools, and systems to support this innovative field.

Holland et al. (2013) discuss the interdisciplinarity inherent in Music Interaction, which they define as the comprehensive process of designing, refining, analysing, and utilising interactive systems involving computer technology for musical activities, emphasising the scientific exploration of this domain (Holland et al., 2013). Collaboration across disciplines – melding the expertise of researchers, designers, and musicians – is often essential. Many contributors assume multiple roles in the creation of Transmutable Music works.

A nuanced evaluation approach addressing each role and developmental phase is imperative. O'Modhrain (2011) offers a framework for assessing

DOI: 10.4324/9781003273554-9

Digital Music Instruments (DMIs) that shares similarities with the context of Transmutable Music products. Her framework, rooted in Human-Computer Interaction (HCI) methodologies, could be adapted for Transmutable Music works, considering the broader scope beyond mere instruments to encompass performance and composition (O'Modhrain, 2011).

In the evaluation model for digital instruments proposed by O'Modhrain (2011), various strategies are recommended for assessing the success of an instrument. These range from multiple stakeholder perspectives, focusing on enjoyment, playability, and robustness to fulfil design specifications.

The model suggests employing critiques, reflective practice, questionnaires, and observational studies to gauge audience enjoyment. To evaluate playability, particularly from a mental model perspective for instrument design, the model advises using experimental approaches.

When considering composers and performers, the recommendation is to utilise reflective practice, cultivate a repertoire, and engage over the long term, potentially extending to longitudinal studies. Quantitative methods are proposed for examining the user interface and its mapping to assess playability for composers and performers.

For the robustness aspect, which pertains to the stability and reliability of the digital instrument, the model calls for quantitative methods to test both hardware and software.

From a designer's standpoint, the evaluation model recommends observation, questionnaires, and informal feedback to measure enjoyment alongside quantitative methods for user interface assessments to determine playability. To verify whether design specifications have been met, designers should consider use cases and feedback related to stakeholder satisfaction.

Lastly, as stakeholders, manufacturers are encouraged to use market surveys and sales data to understand enjoyment levels.

O'Modhrain's model is essential as it highlights the different types of evaluation and roles in the design process. Transferable to Transmutable Music products, composers may work with computer scientists, interaction designers, interface designers, game developers, etc. As much as pilot user studies can aid in some areas of evaluation for composers and producers the most critical evaluation will be evaluating the musical design.

To cultivate an enriching listener or user experience, integrating Music Interaction and HCI strategies is crucial. Music Interaction is seen as a subset of HCI, which, in turn, is a subset of Computer Science, reflecting an academic lineage that includes disciplines such as Electrical Engineering (Holland et al., 2013). However, within the music community, the real measure of an interactive musical system's success is its musicality – its fitness for musical purposes (Holland et al., 2013). Thus, Music Interaction introduces new challenges and methodologies to the field of HCI.

Incorporating evaluation methods from Music Interaction and DMI into developing Transmutable Music products ensures that their designed potential to captivate the listener is achievable. McDermott et al. (2013) suggest that evolutionary and generative music interfaces can gain from HCI and usability research, enhancing their effectiveness and user satisfaction (McDermott et al., 2013). The evaluation framework detailed in this chapter is grounded in such interdisciplinary research, offering a criterion for composers and producers in the Transmutable Music field.

The need for composer and producer evaluation is underscored in 'Interactive Experience in the Digital Age - Evaluating New Art Practice' (Candy and Ferguson, 2014). This article showcases a collection of works where artists and researchers harness evaluation to evolve interactive art, setting a novel research agenda at the intersection of art and technology. The text suggests that HCI and digital arts have become a playground for interdisciplinary work, forging novel interactive art systems and evaluative frameworks and methods (Candy and Ferguson, 2014).

These innovative evaluative strategies are essential, as conventional HCI and product design typically focus on user preferences and experiences without considering the impact on creative practice. Such evaluations extend the frontiers of creativity and contribute to a richer understanding of interactive experiences.

Evaluating Transmutable Music

Transmutable Music, by its very nature, intersects numerous disciplines. To forge Transmutable Music that surpasses the auditory confines of static compositions, it must inherently embody transmutability and variability in its architecture and design.

Chapter 2 delineated the main components of a framework for the musical design of Transmutable Music. This model elaborated throughout the chapter, incorporates key criteria that serve as the basis for the schematic depicted in Figure 7.1. This schematic encapsulates the integration of control systems, content, experience, and musical architecture within Transmutable Music, focusing on two pivotal concepts to underpin the evaluation of a Transmutable Music system: transmutability and variability. During the creative act of composing and producing a Transmutable Music system, these terms are characterised as follows:

- Variability is the myriad of options that allow the music to be altered through the system's musical and sound content and architectural design. It represents the compositional tools at the composer's disposal to craft this potential for change.
- Transmutability is the breadth of the control system's capacity to employ data in transforming and generating content and musical architecture. It encompasses the control system's design within a Transmutable Music system and how data, whether incoming or generated, modulates and reshapes the music.

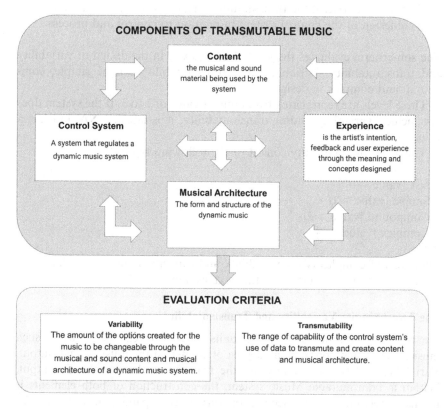

Figure 7.1 Framework for Transmutable Music Design Leading to Evaluation Criteria.

These parameters offer a robust platform that transcends the traditional practices and forms typical in recorded music.

The evaluation criteria emerged during my PhD research, which examined a spectrum of Transmutable Music works and methodologies, including interactive, adaptive, reactive, contextual, algorithmic, and AI-driven processes.

These criteria facilitate the assessment of Transmutable Music, contrasting it with static music forms. They include a set of sub-criteria, which are briefly encapsulated below, providing a comprehensive means to appraise the evolving landscape of music as it shifts from static to dynamic modalities.

Sub-criteria of Variability

1.1 Variability in the musical architecture and content.
1.2 Variability of the resulting form or rendered musical output.

Sub-criteria Transmutability

2.1 The amount of control the transmutability has on the musical architecture.

2.2 The level of control the transmutability has on the musical architecture.
2.3 The design of the control system in relation to control and process.

The sub-criteria evaluates the level of complexity in the design of variability and transmutability. Variability and transmutability can be simple, compound, and complex in design.

These levels are represented by a number value of 1 to 3. If the system does not include variability and/or transmutability, it is labelled NA and has a value of 0.

The levels of complexity consist of the following types:

- NA [value = 0]
- Simple [value = 1]
- Compound [value = 2]
- Complex [value = 3]

The levels of complexity for variability and transmutability are evaluated as per the sub-criteria described in the next section.

The Two Criteria: Variability and Transmutability

The core of Transmutable Music lies in its capacity for change. To assess such music, one should consider the degree of change in the music's final output. Variability and transmutability are the twin forces that shape this output. Within a Transmutable Music system, the construction of both elements is mutually reliant; the architecture of variability expands the horizons for transmutability, and similarly, transmutability informs the design of variability.

Transmutable Music works vary in approach, design, structure and delivery, so the criteria are qualitatively measured. All Transmutable Music works will include these criteria to some degree.

Options for Transmutability: The Transmutability is designed with the intent to infuse the final musical form with variability. It achieves this by utilising data to:

1 Dictate the variability crafted within the musical content and structural design.
2 Forge or reshape the musical content and its structure through various processes.

Choices for Variability: The design of variability in the final form is realised through a two-pronged approach:

1 Employing a suite of compositional techniques to enrich the music with variations, contrasts, branching substructures, layers, blends, and transitional elements.

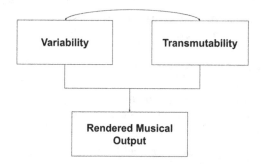

Figure 7.2 The Interdependent Relationship Between Variability and Transmutability.

2 Generating a broader palette of musical and sonic options or expanding the versatility within digital instruments, patches, sound libraries, or procedural audio systems.

The symbiotic relationship between variability and transmutability is the cornerstone of a Transmutable Music piece, as depicted in Figure 7.2. Transmutability is the mechanism that brings designed variability to life. Conversely, without the element of variability, the control system lacks substance to transform, leaving the music unchanged.

The control system manages the variability that has been meticulously composed into the content and structure of the music. The scope of variability that is composed into the system significantly impacts which musical parameters can be data-mapped. These parameters can be manipulated across the work's various hierarchical levels, ensuring that the music is heard and experienced in many ways.

Criterion 1: Variability

In Transmutable Music, variability is vital: compositions are crafted to ensure that each listening instance is a distinct auditory event. This criterion assesses the fluidity of the final musical product and the range of compositional techniques used to achieve such versatility.

Variability is a measurement of the music's capacity for change or its fluidity. Essentially, it is the scale of musical malleability – the degree to which a piece can transform with each performance. Criterion 1(Variability) is concerned with tangible musical elements (like sound and structure) and the overarching design that dictates how these elements interact within a Transmutable Music system. In contrast, Criterion 2 (Transmutability) examines the control mechanisms that guide these transformations.

In order to produce variability in the resulting form, variability must be part of the musical design of a Transmutable Music system. A Transmutable Music system is comprised of four main components: the actual sounds used (content), the structural blueprint (musical architecture), the guiding mechanisms (control system), and the resultant experience. The latter, the experience, is an inherent by-product of how the sound and structure are crafted.

Variability can be evaluated on two levels:

1 Variability in the musical architecture and content
2 Variability of the resulting form or rendered musical output

Sub-criterion 1: Variability in the Musical Architecture and Content

This sub-criterion assesses the breadth and depth of the compositional tools utilised to shape the music's foundational and nuanced elements, as outlined in Chapter 5. These tools are crucial for injecting variability into the sound material and its subsequent layers, broadening the creative horizon with either novel or modified content.

Complexity Levels

Not Applicable (NA)	The sound material or processes are fixed in form and have been composed for a static medium (master audio file)	*Value = 0*
Simple	Variability is present but limited to minor modifications in sound, music content, and mix parameters.	*Value = 1*
Compound	The composition is structured to facilitate variability at the substructural level, influencing the dynamics of instrumentation, harmonics, and thematic structures.	*Value = 2*
Complex	The music is highly dynamic, with content and processes that are in a continuous state of evolution, offering a wide range of variability throughout the piece.	*Value = 3*

Application

- A 'Simple' level might involve basic adjustments like tempo shifts or volume tweaks.
- A 'Compound' level could see the integration of various musical motifs or thematic elements that alter the piece's character.
- A 'Complex' level may utilise AI to introduce new patterns or sounds, creating a piece that is perpetually unique.

Sub-criterion 2: Variability of the Resulting Form or Rendered Musical Output

This sub-criterion is centred on the uniqueness of each listening experience provided by Transmutable Music, ensuring that the musical form is ever-evolving and that no two playbacks are identical.

Complexity Levels

Not Applicable (NA)	The musical experience is consistent across all playbacks, without any variation.	*Value = 0*
Simple	Variations are minimal and subtle, maintaining a familiar structure and musical core.	*Value = 1*
Compound	Some musical attributes are recognisable on each playback - the structure is changed, or some part are changing at relevant hierarchical levels.	*Value = 2*
Complex	Maximum variability is achieved with each playback, presenting a completely new musical discovery, as if hearing the piece for the first time.	*Value = 3*

Application

- A 'Simple' level would offer little to no noticeable change, perhaps a slight variation in intro or outro sequences.
- A 'Compound' level might present alternative bridges or choruses that vary with each session.
- A 'Complex' level could result in a fundamentally different musical journey each time, with new melodies, rhythms, and harmonies.

Criterion 2: Transmutability

Criterion 2 relates to the concept of transmutability – the transformative power of data within a Transmutable Music system. This criterion assesses the extent to which data can alter a piece of music, driven by a control system that interprets and utilises this data to reshape the musical form.

Transmutability is the degree of influence data can have over the evolution of a musical piece. It is about the breadth and depth of change that the data can instigate, guided by the control system's capabilities and its procedural relationship.

The intricacies of a Transmutable Music system's control system are pivotal. They determine how the system interacts with data to influence musical outcomes. For instance, an interactive system would sync user interactions with corresponding musical responses; an adaptive system modifies music in real time to align with gaming scenarios, while a generative system is tasked with creating new sounds and structuring them into a coherent musical architecture.

Every process within the dynamic system defines the kind of data it handles. This is how data is transformed, classified, and ultimately mapped to alter the musical structure. The methodology for how a control system manipulates the musical framework is detailed in Chapters 3 and 5, outlining a model for control system influence at various hierarchical levels.

The purpose behind the processes chosen for a Transmutable Music system is deeply rooted in the artistic vision of the work. This means evaluations focus on the control system's functionality rather than the creative reasons behind its design. The true measure of transmutability is the control system's capacity to change the musical architecture, irrespective of the specific data processes involved.

Enhancing the transmutability aspect of a control system could involve expanding its processes, interactive capabilities, data sets, mapping designs, and the data's transformation and reclassification.

Given the hierarchical nature of Transmutable Music systems, Criterion 2 is broken down into three sub-criteria to address the system's complexity:

1 The level of control the control system has on the musical architecture;
2 The amount of control the control system has on the musical architecture;
3 The design of the control system in relation to control and process.

Sub-criterion 2.1: The Level of Control the Transmutability has on the Musical Architecture

This sub-criterion examines the extent to which the Transmutable Music control system can change the music's structural elements at various levels.

Control Levels

- **Macro Level:** This is the broad, overall structure of the music piece.
- **Meso Level:** This includes the substructures, such as sections and phrases within the piece.
- **Sound Object Level:** The most detailed layer, dealing with individual sounds and notes.

Complexity Assessment

Not Applicable (NA)	No control of the musical architecture.	*Value = 0*
Simple	Control only one level of parametric control to transmute the macro level or only simple transmutability including mixing variability.	*Value = 1*

(*Continued*)

Compound	The control system can manipulate multiple structural parameters, allowing for the alteration of music at the substructural level.	*Value = 2*
Complex	The system offers comprehensive control over all musical parameters, affecting every layer from the overarching structure down to the sound object level.	*Value = 3*

Application

- At a 'Simple' level, the system might adjust the overall volume or transition between major parts of the piece.
- A 'Compound' level would allow for changes in the texture or form within sections, like modifying a chorus or bridge.
- A 'Complex' level implies the ability to generate entirely new melodies or rhythms, effectively rewriting the music during each playback.

Sub-criterion 2.2: The Amount of Control the Transmutability has on the Musical Architecture

Purpose

This sub-criterion quantifies the range of control data exercises over the musical structure within a Transmutable Music system. It determines how many and which types of musical parameters can be dynamically altered by the system's data inputs.

Complexity Levels

Not Applicable (NA)	The system lacks the functionality to modify any musical parameters.	*Value = 0*
Simple	The system can adjust one specific musical parameter, providing a singular dimension of change.	*Value = 1*
Compound	The system has the capability to control multiple musical parameters, allowing for a multi-dimensional transformation of the music.	*Value = 2*
Complex	The system possesses an extensive range of control, capable of altering any and all musical parameters, leading to a highly versatile and dynamic musical output.	*Value = 3*

Application

- To achieve a 'Simple' rating, the system might only change the volume or tempo.
- A 'Compound' rating might involve alterations in pitch, rhythm, and instrumentation.
- A 'Complex' system might integrate algorithmic composition, AI elements, or generative processes, continuously producing new musical variations in real time.

Sub-criterion 2.3: The Design of the Control System in Relation to Control and Process

This sub-criterion assesses the intricacy of the control system's design, particularly in how it processes and interprets data to influence musical output. It examines the system's capacity for real-time data transformation, decision-making, and data-driven creativity.

Complexity Levels

Not Applicable (NA)	The control system doesn't receive or produce data.	*Value = 0*
Simple	The control system is designed to produce and receive data and to transform and map data to musical parameters of precomposed sound material.	*Value = 1*
Compound	The control system is also designed to create and reorganise sound material through some kind of process.	*Value = 2*
Complex	The system employs advanced techniques, such as AI and learning algorithms, to autonomously create sound material and make decisions, reflecting a high degree of sophistication.	*Value = 3*

Application

- A 'Simple' system might adjust predefined elements such as track volume or filter resonance.
- A 'Compound' system could generate variations of musical phrases or rhythms based on user interaction.
- A 'Complex' system might learn from user input to produce entirely new compositions or adaptively reshape music in real time, providing a unique experience with each interaction.

The Framework for the Evaluation of Transmutable Music

A comprehensive evaluation framework has been devised to thoroughly examine the varied methodologies, processes, and structures of Transmutable Music. This framework systematically reviews Transmutable Music works through a multi-step process:

1 Conceptual Overview: This step involves a detailed discussion of the piece, examining its underlying concept, intended goals, objectives, and overall creative approach.
2 Compositional Analysis: Employing the established criteria and sub-criteria, each piece's compositional structure is analysed, illustrating the interplay between the control system, musical content, and overarching architectural design.
3 Criteria Assessment: The work undergoes a qualitative evaluation, offering in-depth insights in relation to the predefined criteria, ensuring a holistic understanding of the piece's attributes and execution.
4 Conclusive Summary: The final step synthesises the findings into a comprehensive summary, encapsulating the essence and efficacy of the Transmutable Music work under scrutiny.

Evaluation Table

The evaluation table is used to summarise the level of complexity of the variability and transmutability in a work's musical design (Table 7.1).

Evaluation Case Studies

Evaluation of Radiohead's **A Moon-Shaped Pool**

A *Moon-Shaped Pool* is an album released by Radiohead. Musically, the album is rich in layering and can be experienced many times to discover

Table 7.1 Evaluation Criteria Table

Evaluation of Transmutable Music	
Level of complexity: NA=0, Simple =1, Compound=2, Complex=3	0 1 2 3
Sub-criteria 1.1 Variability in the musical architecture and content	
A brief justification of the level of complexity	
Sub-criteria 1.2 Variability of the resulting form or rendered musical output	
A brief justification of the level of complexity	
Sub-criteria 2.1 The amount of control the transmutability has on the musical architecture	
A brief justification of the level of complexity	
Sub-criteria 2.2 The level of control the transmutability has on the musical architecture	
A brief justification of the level of complexity	
Sub-criteria 2.3 The design of the control system in relation to control and process	
A brief justification of the level of complexity	

nuances and layers within the recording. The album is evaluated to highlight an instance of a fixed-form album. The choice of artist and album is irrelevant to this evaluation. Any such album format is expected to result in the same evaluation.

Criterion 1: Variability

SUB-CRITERION 1.1: VARIABILITY IN THE MUSICAL DESIGN

The work is fixed in form and has been composed for a static medium (master audio file).

Variability in the musical architecture and content   *0*

SUB-CRITERIA 1.2: VARIABILITY OF THE RESULTING FORM OR MUSICAL OUTPUT

The processes used in the composition of the work do not transmute the resulting music form. The album could be remixed and recontextualised to work within some Transmutable Music work. However, at this time, the composition is fixed and will not work as Transmutable Music. The musical output is static: it is identical on each playback.

Variability of the resulting form or musical output   *0*

Criterion 2: Transmutability

There is no parametric control, only consumer-led transmutability with stop, play and volume controls. Data cannot transmute the musical architecture. The user can only transmute the fixed playback or consumer-led digital transmutations. No data-led digital transmutations are possible in its current form and musical design. The control system does not receive or produce data.

The amount of control the transmutability has on the musical architecture   *0*
The level of control the transmutability has on the musical architecture   *0*
The design of the control system in relation to control and proces   *0*

Summation

Evaluation of Radiohead's A *Moon-Shaped Pool*

This work is not considered Transmutable Music. As shown in the summary table Table 7.2, there is no variability and transmutability of its musical design.

Table 7.2 Summary Table: Evaluation of Radiohead's *A Moon-Shaped Pool*

Evaluation of Transmutable Music	
Level of complexity: NA=0, Simple =1, Compound=2, Complex=3	**0 1 2 3**
Sub-criteria 1.1 Variability in the musical architecture and content	**0**
The work is fixed in form and has been composed for a static medium (master audio file)	
Sub-criteria 1.2 Variability of the resulting form or rendered musical output	**0**
The musical output is static: it is same on each playback	
Sub-criteria 2.1 The amount of control the transmutability has on the musical architecture	**0**
No parametric control, only consumer-led transmutability with stop, play and volume control	
Sub-criteria 2.2 The level of control the transmutability has on the musical architecture	**0**
Data-led transmutability of 0 musical parameters	
Sub-criteria 2.3 The design of the control system in relation to control and process	**0**
There is no control system or interactivity	

Evaluation of 'Is This Your World?'

Conceptual Overview

The proof of concept demonstrates the limitations of stems in a simple system. The single 'Is This Your World?' is used. It was released on my album *Walking Home a Different Way*.

The stems were created to allow a producer to remix alternative versions of the song for release as a B-Side[1] or bonus track.[2] A producer's approach may have been to cut up the existing stems and add effects in addition to adding new instruments or samples.

Criterion 1: Variability

SUB-CRITERION 1.1: VARIABILITY IN THE MUSICAL DESIGN

The work is presented in stems within a fixed song structure; it has been composed for a static medium.

Variability in the musical architecture and content   *0*

SUB-CRITERIA 1.2: VARIABILITY OF THE RESULTING FORM OR
MUSICAL OUTPUT

The processes used in the composition of the work do not transmute the resulting music form except for the panning and gain of each stem. The song could be extended to provide more variability. Music. The musical output is similar: limited variability in playback.

Table 7.3 Summary Table: of Redhead's 'Is This Your World?'

Evaluation of Transmutable Music	
Level of complexity: NA=0, Simple =1, Compound=2, Complex=3	**0 1 2 3**
Sub-criteria 1.1 Variability in the musical architecture and content	1
Limited variability, only gain and panning adjustments.	
Sub-criteria 1.2 Variability of the resulting form or rendered musical output	1
The song doesn't sound that different in each playback	
Sub-criteria 2.1 The amount of control the transmutability has on the musical architecture	1
Limited level of control of musical changes	
Sub-criteria 2.2 The level of control the transmutability has on the musical architecture	1
Only at the master level can the stems be transformed	
Sub-criteria 2.3 The design of the control system in relation to control and process	1
Simple system	

Variability of the resulting form or musical output   *1*

Criterion 2: Transmutability

The only options for the user were to adjust volumes and create different spatial arrangements using panning. Therefore, these basic options limited the audience's ability to create unique song versions.

There is limited parametric control, only consumer-led transmutability with stop, play and volume controls. Data can transmute the musical architecture at the master level. The user can only transmute the consumer-led digital transmutations. Very limited data-led digital transmutations are possible in its current form and musical design. The control system receives minimal data (Table 7.3).

The amount of control the transmutability has on the musical architecture   *1*
The level of control the transmutability has on the musical architecture   *1*
The design of the control system in relation to control and process   *1*

Summation

The composition was not composed for an interactive or transmutable medium. Given the system is simple in design, this experience could be improved by providing more variability options, for example, more layered content (Table 7.3).

The example interface provided a very restricted means to vary and interact with the audio, ultimately limiting creative expression. This was due to a combination of limited interface functionality and limited options within the musical stems, which were originally arranged and mixed for a fixed format. With the inherent limitations in musical content, it can be concluded that there would be limited possibility of interaction and a new approach needed to be explored.

In order to improve this work, more variability needs to be composed to provide more options for the way the song can be played back.

Evaluation of Quarta: Hybrid Game with Generative Soundtrack

The app is a hybrid game with a generative soundtrack developed by game designer Brett J. Gilbert and musician, software designer and Bloom co-creator Peter Chilvers. The game involves playing against an AI-based computer player that learns as you play against it or a multi-player game. This game was chosen to provide an example of a generative music system situated in a hybrid context.

Criterion 1: Variability

SUB-CRITERION 1.1: VARIABILITY IN THE MUSICAL DESIGN

There is limited variability in the sound and music content. There are, however, only two instrument sounds.

Variability in the musical architecture and content   *1*

SUB-CRITERIA 1.2: VARIABILITY OF THE RESULTING FORM OR
MUSICAL OUTPUT

The composition involves piano notes with an ambient string drone backing. The piano has reverb with a depth to it. It builds and falls and uses spatialisation. Although the work is generative and amorphous, its variability is subtle. It starts to sound very similar, with some common melodic patterns appearing occasionally between the lower and higher registers. The musical playback is dynamic and produces changing contexts. However, its limited change in timbre and orchestration means the work sounds fairly similar despite this.

The variability is minimal. Many musical parameters are preserved, changes are limited, and the output sounds similar each time it is experienced.

Variability of the resulting form or musical output   *1*

Criterion 2: Transmutability

Without access to the inner workings of the control system of this app, it would appear to be based on the musical output that the data is transmuting the sound material through a generative system. Data can control two or more levels and parameters. The control system is also designed to create and

Table 7.4 Summary Table: Evaluation of *Quarta*

Evaluation of Transmutable Music	
Level of complexity: NA=0, Simple =1, Compound=2, Complex=3	**0 1 2 3**
Sub-Criteria 1.1 Variability in the musical architecture and content	1
There is limited variability in the sound and music content. There are only two instrument sounds	
Sub-Criteria 1.2 Variability of the resulting form or rendered musical output	1
The variability is minimal; many musical parameters are preserved	
Sub-Criteria 2.1 The amount of control the transmutability has on the musical architecture	2
Control of two or more levels of parametric to transmute (substructures)	
Sub-Criteria 2.2 The level of control the transmutability has on the musical architecture	2
Data-led transmutability on more than one musical parameter	
Sub-Criteria 2.3 The design of the control system in relation to control and process	2
The control system is designed to create and reorganise sound material through a generative process	

reorganise sound material through a generative music process. The control system has a compound [value 2] complexity level.

The amount of control the transmutability has on the musical architecture   *2*
The level of control the transmutability has on the musical architecture   *2*
The design of the control system in relation to control and proces   *2*

Summation

Evaluation of *Quarta*

The control system is more complex than the variability level in the content and musical architecture (Table 7.4). For *Quarta* to increase the variability of its resultant form, it would need to introduce more variability to its musical architecture and sound and musical material.

To conclude, this chapter provides an evaluative framework for the design of Transmutable Music based on the criteria of variability and transmutability.

Summary

Evaluating music, particularly interactive music, requires assessing the music, the user experience, and the system supporting it.

While music interaction research focuses on user experience and system functionality, it often overlooks the musical design's evaluation.

A framework for evaluating Transmutable Music should assess both the musical design and the supportive system's capacity.

Interdisciplinary collaboration involving researchers, designers, and musicians is essential in creating Transmutable Music.

O'Modhrain's evaluation framework for Digital Music Instruments, rooted in HCI methodologies, could be adapted for Transmutable Music. The model highlights different evaluation types and roles in the design process, which is crucial for Transmutable Music involving various collaborators.

Integrating Music Interaction and HCI strategies is key to enriching user experiences and ensuring a system's musicality. Evaluative strategies should consider the impact on creative practice, extending beyond conventional HCI and product design focuses.

This chapter's proposed evaluation framework is grounded in interdisciplinary research, offering criteria for Transmutable Music's composers and producers.

This framework considers the dual concepts of transmutability and variability as central to evaluating Transmutable Music systems.

Variability involves the music's capacity for alteration through the system's content and architectural design.

Transmutability encompasses the control system's ability to transform and generate content and musical architecture using data.

The proposed criteria and sub-criteria provide a structured means to appraise music transitioning from static to dynamic forms.

Sub-criteria of Variability

1.1 Variability in the musical architecture and content.
1.2 Variability of the resulting form or rendered musical output.

Sub-criteria Transmutability

2.1 The amount of control the transmutability has on the musical architecture.
2.2 The level of control the transmutability has on the musical architecture.
2.3 The design of the control system in relation to control and process.

The evaluation case studies illustrate the framework's application, analysing works ranging from fixed compositions to generative music systems.

The case studies show that Transmutable Music requires a balance of variability and transmutability to ensure dynamic and unique listener experiences.

Notes

1 A B-side is the reverse side of a 45-inch single or the less important song on a single of EP. With the A-side being the strongest song.
2 A Bonus track is a piece of music that is added to an edition of remake of an album, single or EP.

References

Candy, Linda, and Sam Ferguson. 2014. "Interactive Experience, Art and Evaluation." In *Interactive Experience in the Digital Age*, 1–10. Cham: Springer International Publishing.

Holland, Simon, Katie Wilkie, Paul Mulholland, and Allan Seago. 2013. "Music Interaction: Understanding Music and Human-Computer Interaction" *Music and Human-Computer Interaction*, edited by Simon Holland, Katie Wilkie, Paul Mulholland, and Allan Seago, 1–28. London Heidelberg New York Dordrecht: Springer.

McDermott, James, Dylan Sherry, and Una-May O'Reilly. 2013. "Evolutionary and Generative Music Informs Music HCI—and Vice Versa." In *Music and Human-Computer Interaction*, 223–240. Berlin/Heidelberg, Germany: Springer.

O'Modhrain, Sile. 2011. "A Framework for the Evaluation of Digital Musical Instruments." *Computer Music Journal* 35 (1): 28–42.

Section 3

Interactive Technologies and Music Making (Background and Context)

8   The Convergence of Remix, Production, and Video Games

The Remix Revolution: Tracing the Evolution of Music's Participatory Culture   ⬩

Music, inherently a dynamic and living art, has long transcended the simple act of listening. Music is participatory. Potts reminds us that music was an active pursuit before the age of recordings – a verb manifesting in the joy of song and live performances. Fast forward to today, and a recorded track has become the default embodiment of a musical piece, relegating live renditions to mere interpretations of these recordings (Potts, 2017). However, this paradigm begs the question – if the music industry began anew today, would we still cling to the static WAV file as our staple commodity?

Music, throughout history, has shown remarkable adaptability, with artists continuously adjusting their creative processes to meet evolving audience demands and leverage technological advancements. Collins offers an academic perspective on the evolution of popular music, noting its adaptability to various playback formats and compositional styles, particularly in response to the temporal demands of film scoring (Collins, 2008a). This adaptability is epitomised in the art of remixing, which technological advancements have profoundly shaped.

Electronic and hip-hop genres have primarily been built upon remixing and sampling existing recorded tracks. This approach traces back to Pierre Schaeffer and his collaborators in 1948, who introduced 'musique concrète,' a concept that treated music as a collage of manipulated audio signals and tape recordings, effectively remixing existing sounds into new creations (Paul, 2003).

As remixing techniques evolved, the Jamaican dancehall scene of the 1960s played a pivotal role, with innovators like King Tubby and Lee Scratch Perry pioneering the 'dub' genre. They manipulated existing tracks to create re-arranged versions with distinctive new rhythms and soundscapes (Arroyo, 2008). These early remixing efforts were intricate and required specialised expertise and costly equipment, marking it as a predominantly producer-driven movement.

However, as technology advanced, the practice of remixing and sampling music became more democratised. This led to a 'street-led' movement, where musicians would release extended tracks for DJs to remix live in clubs.

DOI: 10.4324/9781003273554-11

In the 2000s, an emerging change to music's static form continued. Famous music artists began to release 'stems' – the individual components of a track, like vocals, drums, and bass, intended for use in remixing. This movement echoed the past when artists dropped 12-inch vinyl singles for DJs and enthusiasts to spin their own remixes. The stem releases catered to an increasing consumer appetite for deeper engagement with digital music content.

The iconic band, Nine Inch Nails, was at the forefront, offering numerous tracks in stem format through their remix-centric website, remix.nin.com, which also hosted remixing tools for fans. This initiative was mirrored by other artists such as R.E.M, Radiohead, and Kylie Minogue, who provided stems through various platforms, marking a significant step towards inter-active music by granting fans insight into the production process and inviting them to reimagine it.

Remix competitions gained traction, with platforms like Indaba Music and ccMixter becoming hubs for collaborative and participatory music experi-ences. During this period, Korean Audizen and French Iklax introduced browser-based applications that allowed users to alter song structures and instrumentation – though these companies have since ceased operations (Redhead, 2020).

In a significant development in 2015, Native Instruments unveiled the Stems format, designed for DJs to mix music innovatively. Stems were divided into four components: drums, bass, vocals, and percussion/synths, which were operable via the company's *Traktor Pro 2* software, allowing separate control over each stem's volume and effects. This format was quickly adopted by major dance music labels, revolutionising the DJing experience.

In this digital era, an array of accessible and affordable tools have emerged, enabling anyone with a computer or smartphone to remix and sample music effortlessly, marking a significant shift towards a consumer-led movement.

Diving into the evolution of remix culture reveals a fascinating progression from the era of producer-driven remixes to today's landscape, where fans and users wield the tools to reimagine music. This shift illuminates the historical narrative of artists who continually adapt their methods to align with the evolving demands of their audience.

The digital era has seen the roles of consumers and producers bluring, intensifying the craving for behind-the-scenes access to the creative process. This convergence, highlighted by Bruns (2009), has diluted the once clear boundaries separating the music makers from their audience, placing the music industry at a pivotal juncture of digital engagement (Bruns, 2009).

In the pre-digital age, music's malleability was somewhat constrained. Listeners had a suite of basic controls at their fingertips – play, stop, skip, volume adjustments – but these did not offer the power to reshape the music itself. However, even these modest interactions harboured the seeds of producer-like transmutations, as listeners would craft mixtapes or splice tapes to remix their sonic experiences, stepping into the role of creator.

Despite the strides in technology facilitating remix culture, the future of this practice, especially when intertwined with AI, is complex, mainly due to copyright laws and rights management within a capitalist framework. These legal intricacies can deter bedroom producers from sharing their work, as they risk infringement suits. Lawrence Lessig's Creative Commons sought to address this by fostering an open-source community within software development. Despite Creative Common's aim to simplify licensing, its permanence – once a license is chosen, it is set forever – has led to resistance from some music rights organisations. Nevertheless, this licensing model has found acceptance in the broader open-source community. To cater to the evolving landscape, new start-ups offer standardised licensing options for copyrighted material, with examples like freesound.org, remix competitions, artist-released stems, and services such as Tracklib and Splice that offer accessible and affordable samples.

Production and Electronic Composition

This transformation has been well supported by the rise of Digital Audio Workstations (DAWs) and the evolution of electronic music processes over the past century. Digital Audio Workstations (DAWs) have significantly impacted recorded music's production and composition processes. This is similar to how Audio Middleware software has influenced game audio composition and processes.

Before discussing the details of these changes, it is essential to clarify that we are not referring to a specific genre when refering to electronic music. Instead, we focus on the processes and approaches used to produce music and sound electronically, which can be applied to any genre.

Curtis Roads, an influential figure in this domain, examines how electronic technology has opened new and exciting possibilities for music concerning Edgard Varèse's vision of 'liberation' in 1936 (Varèse & Wen-Chung, 1966). He explains electronic technology's role in freeing music from traditional constraints. One aspect of this liberation is the manipulation of time, where sound can be sped up, slowed down, played backward, or fragmented into tiny pieces, allowing for stretching, shrinking, or scrambling (Roads, 2015).

Moreover, the traditional 12-note equal temperament system for pitch in Western music has been liberated. This liberation has enabled musicians to explore new scales or even abandon scales altogether, leading to sounds that can flow into noise, slow into pulsation, or evaporate and coalesce (Roads, 2015). The advancements in sound synthesis, sampling, and manipulation tools, including software, devices, plugins, libraries, and hardware, have contributed to the liberation of timbre (Roads, 2015).

Spatial audio technologies have also played a significant role in this transformation, liberating space beyond traditional stereo width. New immersive 360-degree binaural, ambisonic, and multi-speaker tools have emerged, expanding the spatial possibilities for music production. *Dolby Atmos*[1] is an ever-expanding format offering an immersive sound experience to consumers.

Electronic music processes can explore infinite sounds, making the laptop a music studio tool with boundless musical possibilities. As a result, the processes of mixing and composition have merged within DAW software. This has provided musicians with the tools to compose the sounds they are working with, whether originating from audio samples or synthesised from oscillators.

In music production within a DAW, composers and producers often simultaneously compose, produce, and mix, representing a significant departure from traditional analogue approaches (Roads, 2015). Despite the newfound creative freedom and infinite possibilities, the DAW alone may not fully cater to artists' needs for working with interactivity.

Fortunately, there are solutions available that offer interactivity within a DAW package. For example, *Max for Live* patches can be integrated into *Ableton Live*, enabling the seamless connection of interactive elements with music production. *Max for Live* and *Max 8* present powerful interactive media programming software that allows musicians to control various parameters, hardware connections, and even physical objects like motors and lights using Arduino and OSC (Bown and Britton, 2013; Ablelton, 2018). *RNBO* is an exciting, novel patching environment of *Max 8.5* that can now export the patches created. It enables users to create web experiences, hardware music devices, audio plugins, and new *Max* objects. Additionally, *RNBO* generates source code, allowing users to utilise it as they see fit.

Moreover, the combination of *Pure Data (Pd)* and *libpd* has provided developers and musicians with an advanced and powerful audio engine for procedural audio, audio processing, and synthesis (Brinkmann, 2012). This interactivity has expanded into gaming middleware audio, where DSP programming languages such as *AudioKit* are poised to replace older frameworks like *Pd* and *Supercollider* (Plans, 2017).

WebPd is a new open-source compiler for the *Pd* audio programming language, allowing users to run.pd patches directly on web pages. Converting audio graphs and processing objects from a patch into human-readable JavaScript or AssemblyScript enables the integration of pure audio-generated code directly into any web application without requiring *WebPd* or *Pure Data again* ('WebPd - Open Collective,' 2023).

The use cases for *WebPd* include publishing generative musical works on the web, executing, tweaking, and collectively sharing patches on any machine, and enabling real-time production, filtering, and playback of sound and music data from web-based applications and games. This technology expands the possibilities for interactive music creation and distribution on the web platform.

Audio middleware software has shifted to cater to composers and sound designers, making audio integration more artistic and uniform (Young, 2013). This trend is also observable in DAW software, where layouts, key commands, and logic flows have become more intuitive based on popular demand (Young, 2013).

Despite these advancements, some barriers to interactive music creation remain, such as the inaccessibility of tools and limited awareness of algorithmic possibilities among certain artists. However, efforts are being made to expand existing tools and improve workflow to overcome these challenges (Brown et al., 2014).

Audio software plugins are now being developed for DAWs and audio middleware software. For instance, iZotope, known primarily for their 'traditional linear mixing, mastering' plugins, has partnered with Audio Kinnect, makers of *Wwise* middleware software. This collaboration allows for real-time processing of audio files, enhancing flexibility and decreasing storage needs (Young, 2013).

With the rise of AI and as these technologies continue to advance, music production has the potential to undergo a complete transformation. Compositions could incorporate procedural, generative, and AI processes, leading to adaptive music that changes and adapts based on data inputs, moving away from the traditional fixed-time music paradigm (Young, 2013).

However, there still needs to be more awareness and understanding about adaptive music in various industries, including game development. Educating audio directors, producers, and composers about the possibilities of interactive music is crucial for its widespread adoption and integration into popular culture and future music products.

Convergence of Music and Gaming Audiences

Music-based video games are emerging as a fascinating nexus where popular music and gaming audiences converge, hinting at the future of entertainment (Collins, 2008b). The evolution of software tools democratises music production, offering new, sophisticated approaches to music creation. Despite extensive documentation of these developments, their impact still needs to be discovered, with the music industry pursuing sustainable revenue models and value creation for their offerings.

The video game market is booming as the music sector struggles to find new income streams for artists. In 2023, the global gaming industry was projected to reach a staggering $US 184.0 billion, with an estimated 3.38 billion individuals worldwide engaging in video games, roughly a third of the planet's population (Newzoo, 2023).

This surge underscores the vast potential of leveraging gaming technology in shaping the future of music. Tessler's analysis delineates the synergistic relationship between the music and gaming industries, positing the potential of video games to succeed in music videos in cultural prominence (Tessler, 2017). She cites Steve Schnur of Electronic Arts (EA), who eloquently compares gaming platforms' historical impact to seminal musical movements and technologies, suggesting that video games may become the cultural equivalents of rock'n'roll and hip-hop for contemporary society (Tessler, 2017).

Despite the critical role music plays within video games, it is only recently that its full potential has begun to be acknowledged. In Japan, where the societal perception of gaming is more inclusive, music from video games has long been promoted to a broader audience. This inclusivity has started to resonate in the West, with an aging gamer population that has begun to appreciate and consume game music as part of the broader cultural milieu (Fritsch, 2013).

The music industry has noted a symbiotic relationship with gaming, using games as platforms to market music and musicians to enhance game sales (Collins, 2008b). Collins underscores that gamers often purchase music discovered within games, indicating the potential for games to serve as a new channel for music consumption (Collins, 2007).

The partnership between the game publisher EA and various record labels exemplifies the economic possibilities at the intersection of music and gaming. In Australia, the collaboration between Remote Control Records and Boss Battle Records aims to amplify the reach and impact of video game composers (Jenke, 2017).

Nonetheless, the integration of music into games has its challenges. Collins (2008b) identifies the inherent 'limited adaptability' of popular music, which is typically produced for fixed playback and lacks the flexibility required for the dynamic environment of games (Collins, 2008a). New strategies, including 'User-generated content,' are being examined to bridge this gap (Collins et al., 2015). Collins provocatively questions whether the structure of popular music might evolve to meet the interactive demands of the gaming industry (Collins, 2008a), encapsulating the broader discussion about the adaptability of music in an increasingly interactive and technologically driven landscape.

Summary

Remixing has transformed from producer-driven to consumer-led movements.

Artists releasing stems encourage fan engagement and personalised music experiences.

Remix culture's history reflects music's adaptability and artists' responses to audience demands.

Electronic music processes have freed music from traditional constraints. Spatial audio technologies and DAWs provide boundless creative possibilities.

The rise of interactive tools within DAW packages like *Max 4 Live* and audio middleware like *Wwise* and *FMOD* offers new dimensions of music production.

The gaming industry's growth presents opportunities for the music industry. Music within games becomes a new channel for consumption and discovery.

The adaptability of music to interactive demands in gaming could hint at changes in popular music structures.

Note

1 Developed by Dolby Laboratories. https://www.dolby.com/technologies/dolby-atmos/

References

Ablelton. 2018. "Max for Live." https://www.ableton.com/en/live/max-for-live/. 2018.

Arroyo, John. 2008. "Evolving the Remix." 2008. http://johnarroyo.com/files/thesis/JohnArroyo-EvolvingTheRemix.pdf.

Bown, Oliver, and Sam Britton. 2013. "Methods for the Flexible Parameterisation of Musical Material in Ableton Live." In *Proceedings of Artificial Intelligence and Interactive Digital Entertainment Conference (AIIDE)*, 24–26.

Brinkmann, Peter. 2012. *Making Musical Apps.* 'O'Reilly Media, Inc.'

Brown, Phil, Rick Cohen, Karen Collins, Mark Estdale, Josh McMahon, Jay Smith, Mikko Suvanto, et al. 2014. "Group Report: Interactive Music Creation and Adoption - The Quest Continues." In *Proceedings of the Nineteenth Annual Interactive Audio Conference PROJECT BAR-B-Q 2014*. https://www.projectbarbq.com/reports/bbq14/Project_Bar-B-Q_2014_report_section_6.pdf.

Bruns, Axel. 2009. "From Prosumer to Produser: Understanding User-Led Content Creation: Towards a Broader Framework for User-Led Content Creation." *Transforming Audiences 2009*. ACM. http://eprints.qut.edu.au/27370/.

Collins, Karen. 2007. 'An Introduction to the Participatory and Non-Linear Aspects of Video Games Audio.' In *Essays on Sound and Vision*, edited by Stan Hawkins and John Richardson, 263–298. Helsinki: University of Helsinki Press. https://pdfs.semanticscholar.org/b6df/39d9d90f82f25c1ce8a78cdbe52dfbdcdf34.pdf.

Collins, Karen. 2008a. *From Pac-Man to Pop Music: Interactive Audio in Games and New Media*. Farnham, United Kingdom: Ashgate Publishing Ltd.

Collins, Karen. 2008b. *Game Sound: An Introduction to the History, Theory, and Practice of Video Game Music and Sound Design*. Cambridge, MA: The MIT Press.

Collins, Karen, Alexander Hodge, Ruth Dockwray, and Bill Karpalos. 2015. "Veemix: Integration of Musical User-Generated Content in Games." In *Proceedings of the Audio Engineering Society Conference: 56th International Conference: Audio for Games*. Audio Engineering Society.

Fritsch, Melanie. 2013. "History of Video Game Music." In *Music and Game*, 11–40. Berlin/Heidelberg, Germany: Springer.

Jenke, Tyler. 2017. "EXCLUSIVE: Remote Control Inks Deal with Video Game Music Label Boss Battle." The Industry Observer. 2017. https://www.theindustryobserver.com.au/remote-control-announces-partnership-boss-battle-records/.

Newzoo. 2023. "Global Games Market Report: An Overview of Trends and Insights."

Paul, C. 2003. *Digital Art*. London: Thames Hudson.

Plans, Elise. 2017. "Composer in Your Pocket: Procedural Music in Mobile Devices." *Music on Screen* 51.

Potts, S. 2017. "Why the Music Industry Needs Computer Coders." 2017. http://www.hypebot.com/hypebot/2017/07/why-the-music-industry-needs-computer-coders-.html.

Redhead, Tracy. 2020. "Dynamic Music: The Implications of Interactive Technologies on Popular Music Making." NSW, Australia: University of Newcastle.

Roads, Curtis. 2015. *Composing Electronic Music: A New Aesthetic*. USA: Oxford University Press.

Tessler, Holly. 2017. "The New MTV? Electronic Arts and'playing'music." In *From Pac-Man to Pop Music*, 13–25. London, United Kingdom: Routledge.

Varèse, Edgard, and Chou Wen-Chung. 1966. 'The Liberation of Sound.' *Perspectives of New Music* 5 (1): 11–19. 10.2307/832385.

"WebPd - Open Collective." 2023. https://opencollective.com/webpd.

Young, David M. 2013. "The Future of Adaptive Game Music: The Continuing Evolution of Dynamic Music Systems in Video Games." *In Proceedings of the AES 49th International Conference*, 6–9.

9 The Convergence of Music, Technology, and Interactivity

Interactive Art and Composition

This chapter briefly examines the technologies that have shaped the domain where the listener's interaction and engagement take centre – first, delving into Interactive Art and Composition.

Although interactive technologies are changing and influencing society, it is not just technology that inspires our desire to explore concepts of interactivity, choice and control.

Interactive art transcends the notion of passive observation, inviting active engagement from the audience, which has a long and rich history documented. Since the early 20th century, artists have been actively seeking to draw the viewer into the creative process, challenging the conventional constraints of art (Kwastek, 2015). Interactive art is inherently multidisciplinary, blending visual, time-based, and performance arts into a cohesive experience (Kwastek, 2015).

The historical trajectory of interactive sound installations and the design of novel musical instruments from the 1970s onwards has been chronicled by scholars like Drummond (2009), who underscores the profound influence these early endeavours have had on shaping interactive music today.

The Ars Electronica Festival in Linz stands as a testament to this evolution, showcasing a diverse array of interactive artworks since 1990. These works provide valuable insights into the development of interactive technologies over the past few decades, offering a window into how these innovations have infused contemporary music production, particularly in the digital age where integration with visual arts, gaming, and design is increasingly prevalent.

Pioneering sound artists have been forging paths in interactive media since the 1960s. These early adoptions of interactive elements have laid the groundwork for composers to push the boundaries of traditional music composition (Sweet, 2015). Even before the 20th century, as explored in Chapter 6, composers explored the possibilities of chance, aleatoric, and indeterminate music to infuse variability into performances. However, a formal system within popular song-based music still needs to be discovered. That said, new developments introducing randomisation in music production

DOI: 10.4324/9781003273554-12

DAW software have seen more artists utilising these processes in compositional practice. Instead, contemporary practices often draw from the gaming industry and generative and adaptive music concepts.

Joel Chadabe, a seminal figure in this field, introduced the term 'interactive composition' in 1967, encapsulating a method that involves creating a real-time, performable computer music system and actively composing and performing with it. His innovations paved the way for dynamic compositions that respond in real-time to performers' actions, effectively merging the roles of composer and performer (Chadabe, 1984).

Chadabe (1984) delineates the intricate process of interactive composition, which unfolds in two critical stages:

1 The design of a bespoke interactive system for composition.
2 The simultaneous act of composing and live interaction with the system.

Creating such a system entails assembling a programmable computer, a synthesiser, and at least one interface for performance. The programming logic of the computer is crafted to function autonomously and in real-time, with the capabilities to:

- Decode the performer's actions, turning them into partial controls for the unfolding musical narrative.
- Fabricate control mechanisms for musical elements not directly influenced by the performer.
- Command the synthesiser to produce the desired sounds.
- This process underscores the importance of establishing a framework that supports interactive artistry and functions similarly to traditional music control systems, like musical notation (Chadabe, 1984).

Brian Eno, speaking in 1995, projected a future where the music marketplace would shift from selling discrete musical compositions to offering malleable systems that empower listeners to personalise their auditory experiences. Eno envisaged a creative space where artistry, scientific exploration, and playful interaction converge, heralding a new era where artists continually innovate within this intersection (Kelly, 1995).

Eno's insights were prescient when interactive CD-ROM music projects were emerging. Over the subsequent quarter-century, artists have increasingly embraced and explored these concepts, leading to the development of Transmutable Music offerings that are shaped by listener interaction and engagement.

Music Video Games

Music video games such as Guitar Hero, SingStar, and Rock Band have brought music and gaming enthusiasts together and cultivated a new wave of musically inclined gamers who engage with music more artistically than before (Peerdeman, 2010). These games fall into a distinct niche where musical interaction is central, unlike traditional video games where music typically serves as an atmospheric backdrop (Pichlmair and Kayali, 2007).

Pichlmair and Kayali categorise music video games into rhythm games and instrument games, with notable mentions of audio art like Fijuu by Julian Oliver and Masaki Fujihata's Small Fish, which exist outside traditional gaming industry channels. Their seminal paper 'Level of Sound' (2007) identifies seven key features defining these music video games: active scores, rhythm action, quantisation, synaesthesia, play as performance, free-form play, and sound agents (Pichlmair and Kayali, 2007).

Liebe (2013) provides a concise breakdown of these concepts, explaining quantisation as a mode that simplifies music alignment for players, thereby easing musical expression within a predefined structure. Sound agents are interactive elements with inherent behaviours that produce music upon engagement. Rhythm action challenges players to match sequences to score points, while active scores allow for the adaptability of music scores in performance. Free-form play is associated with digital toys rather than structured games, offering an open-ended mode of play found in various music games. Synaesthesia blends visual and auditory stimuli, enhancing player immersion and play as performance emphasises the player's physical interaction as part of the gaming experience (Pichlmair and Kayali, 2007; Liebe, 2013).

Collins (2008) further describes music as the driving force in three game categories: music-themed games featuring bands or artists, creative games, and rhythm-action games. Notably, the band Journey pioneered in this respect, being the first band featured in a video game in 1982. The Atari game by Data Age was called *Journey's Escape*. Collins (2008) describes the game, which came out as an arcade game a few months later. She notes the game 'relied on built-in sound chips, but during a special bonus concert scene, a hidden cassette player inside the arcade machine would play their hit Separate Ways' (Collins 2008, 112). Many famous music artists have since released musician-themed games, including Michael Jackson in 1990 and Aerosmith in 1994.

Webster (2009) cites *Dance Studio* as the first music video game, dating back to 1987, where players mimicked a dance instructor's moves on a NES Power Pad (Webster, 2009). However, *Otocky* and *To Be on Top*, both released the same year, are also recognised as early music video game innovations, arriving just years after *Journey's Escape*.

PaRappa the Rapper is another hallmark in the genre, with on-screen cues for rhythm matching that furthered the genre's success in Japan. Its co-developer Matsuura, who transitioned from a music to a gaming career, expressed in an interview with Webster the desire to shape the medium to realise his musical vision, underscoring the collaborative nature of musical creation in gaming (Webster, 2009).

Collins (2008) examines rhythm-based games like *Britney's Dance Beat*, which incorporated the act of auditioning for a dance tour with Britney Spears. The gameplay was closely tied to the timing of button presses, with the music fixed to Britney's recorded tracks (Collins 2008).

The gaming industry witnessed a surge in the popularity of remix and karaoke-style games, with titles like FreQuency and SingStar leading the charge. Harmonix's FreQuency, in particular, assembled a roster of 27 renowned music acts spanning various genres – from electronic music maestros to celebrated DJs and mainstream pop groups. Among the high-profile names that graced the game's soundtrack were The Crystal Method, No Doubt, Paul Oakenfold, and Orbit. FreQuency encountered marketing challenges despite its innovative concept due to its complex visuals, as noted by one of the game's creators in a Gamasutra interview (Alexander, 2004). Harmonix continued to innovate with Amplitude, showcasing talents like Garbage, Herbie Hancock, and Pink. Notably, these games incorporated remixes of popular tracks tailored for gameplay rather than original compositions designed for the gaming environment. Financially, it was an arduous journey for the company, which faced losses on these ventures over a decade. Nevertheless, Harmonix's creative spirit was undeterred, paving the way for its monumental success with the Guitar Hero and Rock Band franchises.

The influence of music video games on the industry extends to innovative titles like Electroplankton and Tenori-On, which have significantly impacted game audio. Additionally, the alternative reality game Year Zero, based on the Nine Inch Nails concept album, represents a groundbreaking fusion of music and gaming, described by Trent Reznor as a novel form of entertainment that transcends mere marketing gimmicks (Wikipedia, 2011).

While this discussion centres on popular music within video games, it is essential to recognise the broader landscape of pioneering and avant-garde music games that also contribute to the evolution of this interactive art form.

Interactive CD ROM

In the dawn of the 1990s, the fusion of music and gaming took an innovative turn with the advent of CD-ROM technology, offering music artists a new canvas to blend their art with interactive gaming concepts. The CD-ROM format broke ground by enabling the inclusion of lengthy recorded audio tracks in games, a significant leap from the constraints of wavetable synthesis and FM synthesis techniques prevalent at the time. Collins (2008) noted this profound impact on composers and sound designers, allowing them to reliably predict how their audio would perform across standard consumer setups and incorporate a richer palette of sounds, including live-recorded effects, instruments, vocals, and in-game dialogues (Collins, 2008).

This technological milestone sparked what Young (2010) calls a mini-boom, with artists venturing into CD-ROM projects that pushed the boundaries of interactive audio within the limitations of the era's technology (Young, 2010).

Noteworthy examples from 1994 include Peter Gabriel's *Xploral*, Prince's *Interactive*, Laurie Anderson's *Puppet Motel*, and David Bowie's *Jump*, which combined cutting-edge 3D visuals with the music to create immersive experiences.

The year 1997 saw the group Coldcut release *Let Us Play*, an album bundled with the Hex CD-ROM and Playtools. This interactive suite offered a new degree of engagement, featuring a collection of digital games, toys, and a selection of the band's music videos, which users could actively remix through the included VJAMM sequencing programme (Stilwell, 2017).

Playtools itself was a creative playground for audio experimentation. It provided users with sliders to manipulate sound across three banks – drum loops, bass loops, and various sonic textures. Users could interactively alter the mix, creating a dynamic electronic soundscape that kept the audio experience fresh and engaging (Pepperell, 2002).

This interactive approach was inspired by *Generator*, an innovative installation showcased at the Glasgow Gallery of Modern Art, which Pepperell (2002) describes as featuring two consoles for live manipulation of audio and visuals, enabling even those without musical expertise to orchestrate eclectic soundscapes combining diverse genres like rock, opera, and hip-hop (Pepperell, 2002).

The success of *Generator* led to the creation of Playtools and paved the way for further innovative projects. These developments formed the foundation for Coldcut's *Let Us Play* and *Let Us Replay* packages, which Stilwell (2017) praises as a significant divergence from conventional music media, transforming passive consumers into active participants in the music experience (Stilwell, 2017). Coldcut also brought us the successful international electronic and dance label Ninja Tune[1] and has continued to work in this space, releasing *Ninja Jamm*[2] and the new *Jamm Pro*.[3] This shows the band's commitment to remix culture and new technologies. They have proved to be one of the innovators in this field.

Young (2010) highlights that these forays into CD-ROMs often included videos and interactive features that allowed users to manipulate the content (Young, 2010). However, the high cost of production and a lack of sustained consumer interest, coupled with the rapid evolution of web technologies, meant that the CD-ROM's era as a trailblazer for interactive music was fleeting, ultimately overtaken by the rise of the internet as a more dynamic and accessible platform for music distribution and engagement.

The Web – Browser-Based Audio

The internet became a fertile ground for musical innovation, offering music artists an expansive new medium to craft interactive experiences. With the release of Adobe Flash 5 in 2000, equipped with ActionScript, it became possible for creators to design far more complex and engaging user interfaces (Warren, 2012).

In a pioneering move in 2001, the BBC and Fat Boy Slim collaborated on an Adobe Flash-based remix game, inviting users to create their own dance floor vibes by mixing tracks directly in their web browsers (BBC, 2001; Fat-Boy-Slim, 2004). Users could manipulate loops of drums, effects, vocals, and samples across a mixing interface, although the tool's simplicity meant that mixes were only sometimes seamless.

Despite the rapid pace of technological obsolescence, some Adobe Flash-based music projects like Incredibox[4] have endured, allowing users to arrange music using animated characters with substantial online engagement, boasting millions of recorded mixes and a significant following since its inception in 2009.

The advance of technology led to the end of Adobe Flash in 2020, as the industry shifted towards universal web standards such as HTML5, WebGL, and WebAssembly (AdobeCommunications, 2017). These open standards obviate the need for plugins like Flash, streamlining the development of web-based musical applications.

Integrating the Web Audio API across browsers has revolutionised the creation and manipulation of sound in web applications, enabling dynamic creation and modification of audio, sophisticated visualisations of audio data, and even analysis for beat detection or instrument identification (Pfeiffer and Green, 2015). This API supports intricate operations like multi-track playback and the addition of synthetic effects, as demonstrated by platforms like Chrome Music Lab, which showcases the creative potential of the available Web Audio API tools on GitHub.

Browser-based musical applications have achieved cross-platform functionality, potentially operating across desktop and mobile devices. Roberts et al. (2013) emphasise the browser's unique position as a universal runtime environment, facilitating the creation of 'write once, run anywhere' musical interfaces that can leverage device-specific features like accelerometers, multi-touch interfaces, gyroscopes, and advanced sound synthesis APIs (Roberts et al., 2013).

Despite these advances, the music industry has hesitated to adopt browser-based music as a significant revenue stream, perhaps due to the historical challenges with monetisation and cross-browser compatibility. However, as mobile technology continues to dominate and browsers grow increasingly capable, the opportunities for innovative web-based music experiences are vast and ripe for exploration.

Mobile Devices and Apps

Mobile devices have developed into compact multi-sensory computers, introducing novel ways of interaction, such as multi-touch gestures, accelerometer control, or geolocation tracking, into everyone's hands and pockets (Thalmann et al., 2016; Essl and Rohs, 2009). Their wide-ranging sensor capabilities allow for interactivity in live and recorded music, influencing a range of new works and potential for new music forms and interaction in the future.

The advent of mobile technology has cast its influence on music, giving rise to the field of 'Mobile Music,' introduced by Gaye et al., 2006, which focuses on musical interaction in mobile contexts using portable technology (Gaye et al., 2006). Artists have leveraged mobile phones as instruments, pioneering the use of this technology in musical creation (Essl and Rohs, 2009; Geiger, 2003; Schiemer and Havryliv, 2006; Tanaka, 2004). This innovation has prompted popular music artists to explore new formats, integrating mobile technology into their artistry (Dibben, 2014). Popular music artists began releasing apps in 2009 (Dibben, 2014; Dredge, 2016), though many early examples were used primarily as promotional tools for an album release, similar to a band's website.

'Custom-made apps are expensive; therefore, many artist apps are instead based on a toolkit comprising a selection from different modules, including free streaming services for an artist's audiovisual content, photographs, news, chat rooms, and the option to buy music, tickets, and other merchandise' (Dibben, 2014, 683). Geere (2011), in his *WIRED* article, critiques the early band-related apps as being mostly lacklustre and redundant due to their similarity to existing band websites. He argues using apps as standalone art forms rather than mere marketing tools. He suggests the potential for apps to use contextual data to influence the music, leveraging the extensive capabilities and sensors available on a mobile device (Geere, 2011).

There are various types of music apps, such as promotional apps, remix apps, instrument and sequencer apps, sing-and-play-along apps, in-music-action game apps, and new audiovisual music album releases, which have seen a wealth of experimental music apps over the past decade (Dibben, 2014).

RjDj was a trailblazer in the personalisation of music through mobile apps, founded in 2009 by Michael Breidenbruecker. Breidenbruecker explains that while Last.fm personalised playlists, RjDj personalised the song itself, a breakthrough made possible by recent technological advances (Nordgren, 2009). A range of popular music artists produced mobile apps with 'reactive music,' which, rather than fixed playback and form, uses input data to generate music that adapts to the listener's environment by employing the built-in sensors of mobile devices (Bauer and Waldner, 2013). Waldner et al. (2011) note that by the end of 2010, RjDj had achieved over 15 million distributed 'reactive' music works and 3 million downloads, exemplifying the potential of music distribution as software (Waldner et al., 2011). Unfortunately, RjDj was ahead of its time and is no longer available.

Album apps have been diversifying since 2009, with interactive elements becoming more common, though many still feature static music. The format of what constitutes an album app varies, with some offering expansive interactive experiences and others serving as digital collections of songs and artistic content (Dredge, 2016; Toulson et al., 2016). This innovation has paved the way for music to be transformed, with companies like RjDj and influential works by artists and developers like Scott Snibbe and Björk pushing the boundaries of what an album app can be (Shakhovskoy and Toulson, 2015).

Some examples of album apps include Sting 25, which is an 'umbrella app for Sting's entire back catalogue' (Dredge, 2016), and Kim Gun Mo's karaoke app. Although Little Boots' album app contained only four tracks, it was still considered an 'album app.' An album app associates an app with a traditional album, even if it is nothing like one.

Toulson, Paterson et al. provide a good definition, 'The album app format is valuable since it allows unique artistic and interactive content to be distributed alongside a collation of audio, supporting the notion that an album is more than just a collection of songs, but potentially a representation of artistic vision which may include artwork, photography, lyrics, video, animation, gaming, social networking and crucially - interaction' (Toulson et al., 2016). With this definition, many of the works listed in Table 9.1 might be considered album apps. Over time, more apps are being developed that involve transforming the actual music itself.

RjDj were influential because of their first popular music release in 2009, *Little Boots*. Scott Snibbe, Björk's collaborator on the *Biophilia* app, was also

Table 9.1 Overview of some of my favourite mobile Transmutable Music apps

Work	Authors	Year	Playback	Audio System
Bloom	Brian Eno/Peer Chilvers	2008	Mobile App	Generative/ interactive
Air	Brian Eno/Peer Chilvers	2009	Mobile App	Generative
Little Boots Reactive Remixer	Little Boots 2009	2009	Mobile App	Interactive/ Reactive
Love	Air 2010	2010	Mobile App	Interactive/ Reactive
The National Mall	Bluebrain	2011	Mobile App	Contextual
Biophillia	Bjork	2011	Mobile App	Interactive
Ninja Jamm	Ninja Tunes	2012	Mobile and Tablet App	Interactive
Yellofier	Boris Blank and Hakan Lidbo	2013	Mobile and Tablet App	Interactive
WEAV	Lars Rasmussen	2015	Mobile App and Audio Software	Contextual
Spotify Running	Spotify	2015	Mobile App	Contextual
Red Planet	Daisy and the Dark	2016	Mobile App	Interactive
Fantom	Massive Attack	2016	Mobile App	Interactive/ Contextual
Nagaul Sense	Nagaul Sounds	2016	Mobile App	Interactive
Mubert	Mubert	2016	Mobile App	Autonomous/ Algorithmic

influential, producing some exciting and cutting-edge album apps, *Metric Synthetica* and the Philip Glass album *Rework*.

Piano Ombre By Francois and the Atlas Mountains is an exciting album app that Shakhovskoy and Toulson (2015) discussed in detail. Interestingly, it was the first app that was eligible to chart – potentially providing a framework for other 'album apps' to be chart-worthy, an essential marker for success in the music industry. The app was 'designed to provide a digital music album experience that maintains the rich visual media and additional artistic content that would be expected in an analogue sleeve, and much more as well. The listener can play the album while browsing song lyrics, production credits, photographs, biographies, artwork, animations, and can also access exclusive studio outtakes and B-side tracks' (Shakhovskoy and Toulson, 2015). The music presented, however, remained the same as a static album. It is an interesting example; will new generations even consider owning an album app to engage with their favourite artists? Especially when they can use social media and the internet to engage easily.

Within this context, essential album apps involve a combination of audio, visual, and interactive elements that go beyond the traditional album format. I have noted some of my favourite apps in Table 9.1 that have been released that showcase innovative ways of using Transmutable music.

This is an exciting time to observe the progress of the music industry. It is evident how technological advancements have brought us to this point where it is impossible to predict where Transmutable Music's products and services might lead us, especially with the integration of AI, Blockchain, and future technologies.

Summary

The chapter discussesthe influence of interactive technologies on society and the longstanding tradition of interactive art, which invites active audience participation beyond passive observation.

Scholars document the historical progression of interactive sound installations and novel musical instruments starting in the 1970s, significantly shaping today's interactive music landscape.

The Ars Electronica Festival in Linz has been an important platform for showcasing interactive artworks since 1990, reflecting the advancements in interactive technologies and their incorporation into modern music production.

Joel Chadabe introduced 'interactive composition' in 1967, a methodology involving creating real-time, performable computer music systems and engaging with these systems in active composition and performance.

Brian Eno predicted a shift in the music marketplace towards systems that allow listeners to personalise their auditory experiences. This prediction has materialised over the past 25 years with the development of Transmutable Music.

Music video games like Guitar Hero and SingStar have merged music and gaming communities, offering musically inclined gamers more artistic engagement with music.

The internet has served as a fertile ground for musical innovation, with Adobe Flash enabling the creation of complex user interfaces and the Web Audio API revolutionising audio creation and manipulation in web applications.

Mobile devices have evolved into compact computers, providing new ways to interact with music and giving rise to the field of 'Mobile Music,' which focuses on musical interaction using portable technology.

Artists have utilised mobile phones as instruments, leading to the release of apps by famous music artists that integrate mobile technology into their artistic expression.

The chapter highlights RjDj's contribution to personalising music through 'reactive music' mobile apps. This novel approach uses built-in sensors on mobile devices to generate music that adapts to the listener's environment.

Album apps have diversified since 2009, with interactive elements becoming increasingly prevalent, offering a range of interactive experiences that extend beyond the traditional album format.

Notes

1 https://www.ninjatune.net/home
2 https://www.ninjajamm.com/
3 https://jammpro.net/
4 https://www.incredibox.com/

References

AdobeCommunications. 2017. "Flash & The Future of Interactive Content." https://theblog.adobe.com/adobe-flash-update/: Adobe. 2017.

Alexander, L. 2004. "Amid a Struggling Kickstarter, Harmonix Reflects on Amplitude." https://www.gamasutra.com/view/news/217994/Amid_a_struggling_Kickstarter_Harmonix_reflects_on_Amplitude.php. 2004.

Bauer, Christine, and Florian Waldner. 2013. "Reactive Music: When User Behavior Affects Sounds in Real-Time." In *Proceedings of CHI'13 Extended Abstracts on Human Factors in Computing Systems*, 739–744. ACM.

BBC. 2001. "Play the Fatboy Slim Online Mixing Game." http://www.bbc.co.uk/radio1/ibiza2001/fatboy_game.shtml. 2001.

Chadabe, Joel. 1984. "Interactive Composing: An Overview." *Computer Music Journal* 8 (1): 22–27. 10.2307/3679894.

Collins, Karen. 2008. *Game Sound: An Introduction to the History, Theory, and Practice of Video Game Music and Sound Design*. Cambridge, MA: The MIT Press.

Dibben, Nicola. 2014. "Visualizing the App Album with Björk's Biophilia." In *The Oxford Handbook of Sound and Image in Digital Media*, edited by C. Vernallis, A. Herzog, and J. Richardson, 682–706. Oxford: Oxford University Press. 10.1093/oxfordhb/9780199757640.013.01.

Dredge, Stuart. 2016. "The History of the Album-App: Creativity Galore but Commercially Tough." http://musically.com/2016/07/25/history-album-app-creativity-commercially/: Music Ally. 2016.

Drummond, Jon. (2009). "Understanding Interactive Systems." *Organised Sound* 14 (2): 124–133.

Essl, G., and M. Rohs. 2009. "Interactivity for Mobile Music Making." *Organised Sound* 14 (2): 197–207.

Fat-Boy-Slim. 2004. "Online Mixing Game." BBC Radio 1. 2004. http://www.bbc.co.uk/radio1/fbs/pcexplorer/fbsexppc.html.

Gaye, Lalya, Lars Erik Holmquist, Frauke Behrendt, and Atau Tanaka. 2006. "Mobile Music Technology: Report on an Emerging Community." In *Proceedings of the 2006 Conference on New Interfaces for Musical Expression*, 22–25. IRCAM—Centre Pompidou.

Geere, Duncan. 2011. "It's Time for Mobile Music Apps to Grow Up." https://www.wired.com/2011/04/mobile-music-apps/: WIRED. 2011.

Geiger, Günter. 2003. "PDa: Real Time Signal Processing and Sound Generation on Handheld Devices." In *ICMC*.

Kelly, Kevin. 1995. "Gossip Is Philosophy." https://www.wired.com/1995/05/eno-2/?pg=4&topic=. 1995. https://www.wired.com/1995/05/eno-2/?pg=4&topic=.

Kwastek, Katja. 2015. *Aesthetics of Interaction in Digital Art*. Cambridge, MA: The MIT Press.

Liebe, Michael. 2013. "Interactivity and Music in Computer Games." In *Music and Game*, 41–62. Berlin/Heidelberg, Germany: Springer.

Nordgren, A. 2009. "Epilogue Reactive Music and Invisible Interfaces." *IPhone User Interface Design Projects*, 235–238.

Peerdeman, P. 2010. "Sound and Music in Games." VU.

Pepperell, Robert. 2002. "Computer Aided Creativity: Practical Experience and Theoretical Concerns." In *Proceedings of the 4th Conference on Creativity & Cognition*, 50–56. ACM.

Pfeiffer, Silvia, and Tom Green. 2015. *Beginning HTML5 Media: Make the Most of the New Video and Audio Standards for the Web*. New York, NY: Apress.

Pichlmair, Martin, and Fares Kayali. 2007. "Levels of Sound: On the Principles of Interactivity in Music Video Games." In *Proceeding of DiGRA Conference*. Citeseer.

Roberts, Charles, Graham Wakefield, and Matthew Wright. 2013. "The Web Browser as Synthesizer and Interface." In *Proceedings of NIME*, 313–318. Citeseer.

Schiemer, Greg, and Mark Havryliv. 2006. "Pocket Gamelan: Tuneable Trajectories for Flying Sources in Mandala 3 and Mandala 4." In *Proceedings of the 2006 Conference on New Interfaces for Musical Expression*, 37–42. IRCAM—Centre Pompidou.

Shakhovskoy, Jonathan, and Rob Toulson. 2015. "Future Music Formats: Evaluating the 'Album App.'" *Journal on the Art of Record Production*, no. 10.

Stilwell, Robynn. 2017. *Changing Tunes: The Use of Pre-Existing Music in Film*. London, UK: Routledge.

Sweet, M. 2015. *Writing Interactive Music for Video Games: A Composers Guide*. US: Pearson Education Inc.

Tanaka, Atau. 2004. "Mobile Music Making." In *Proceedings of the 2004 Conference on New Interfaces for Musical Expression*, 154–156. National University of Singapore.

Thalmann, F., A. P. Carrillo, G. Fazekas, G. A. Wiggins, and M. Sandler. 2016. "The Mobile Audio Ontology: Experiencing Dynamic Music Objects on Mobile Devices." In *2016 IEEE Tenth International Conference on Semantic Computing (ICSC)*, 47–54. 10.1109/ICSC.2016.61.

Toulson, R., J. Paterson, S. Lever, T. Webster, S. Massey, J. Ritter, R. Hepworth-Sawyer, J. Hodgson, R. Toulson, and J. Paterson. 2016. "Interactive Digital Music: Enhancing Listener Engagement with Commercial Music." In *Innovation in Music II*. UK: Future Technology Press.

Waldner, Florian, Martin Zsifkovits, Lana Lauren, and Kurt Heidenberger. 2011. "Cross-Industry Innovation: The Transfer of a Service-Based Business Model from the Video Game Industry to the Music Industry." In *Proceedings of International Conference on Emerging Intelligent Data and Web Technologies.* 10.1109/eidwt.2 011.30.

Warren, C. 2012. "History of Flash." https://mashable.com/2012/11/19/history-of-flash/#GzTGOGzZwPq8. 2012.

Webster, A. 2009. "Roots of Rhythm: A Brief History of the Music Game Genre." https://arstechnica.com/gaming/2009/03/ne-music-game-feature/. 2009.

Wikipedia. 2011. "Year Zero (Video Game)." 2011. https://en.wikipedia.org/wiki/Year_Zero_(video_game).

Young, D. 2010. "MUSICAL CD-ROMS." http://www.inventinginteractive.com/2010/10/11/musical-cd-roms/). 2010.

Section 4

Compositional Approaches in Transmutable Music

Tutorial Series

10 Mobile Music Making Basic Tutorials (MIDI)

Overview of Section 4

This section presents eleven tutorials and two practical project tasks, with several available on the Companion Website for a dynamic learning experience. Each tutorial should take around 2 hours. Transmutable Music encompasses multiple disciplines, including music composition, production, computer science, data science, interaction design, and user experience. The tutorials are crafted to be accessible, avoiding the need for in-depth knowledge in any single area.

In what follows is an outline of the tutorials in this section.

Chapter 10 Mobile Music Making: Basic Mapping Tutorials (MIDI)

10.1 Working with *TouchOSC*
10.2 Simple Mapping to Musical Parameters (MIDI)
10.3 Designing a Template/GUI with TouchOSC (MIDI)

Chapter 11 Mobile Music Making Advanced Tutorials (OSC)

11.1 Accessing Accelerometer data in Max4Live (OSC) – working with LOM
11.2 Developing Simple Models with Decision Trees M4L (OSC)
11.3 Intro to Machine Learning using Accelerometer Data in M4L
Creative Exercise - Crafting Your Interactive Music Prototype with Mobile Integration

Chapter 12 Adaptive Music for Video Games

12.1 Vertical Orchestration (Layering and Blending): Building Suspense and Intensity.
12.2 Prototyping and Audio Integration: Intro to FMOD and Unity
12.3 Sound Effects and Variability in Unity: the Viking Village Project
12.4 Horizontal Resequencing for Adaptive Audio: FMOD
12.5 Horizontal Resequencing and Audio Integration: FMOD and Unity
Creative Exercise - Composing Adaptive Audio and FX for the *Viking Village*
These serve as a springboard from which you can further specialise and deepen your expertise. The skills, approaches, and methods taught, while

DOI: 10.4324/9781003273554-14

demonstrated in specific software, are designed to be transferable to other platforms and applications.

These tutorials aim to establish a foundational knowledge base that you can build upon. To facilitate this, I have crafted project ideas for you to implement the skills you've learned. As we progress through each tutorial, I will provide resources to enhance your understanding in many vital areas.

Mobile Music Making Series – Chapters 10 and 11

This tutorial series guides you through various practical examples. It includes working with data, exploring compositional approaches, utilising controllers, developing graphical user interfaces (GUIs), interaction design, assessing user experience, and applying simple to complex parameter mapping techniques within *Ableton Live*. We will also cover data transformation and mapping and explore machine learning for interpreting accelerometer data.

We will be creating a mobile music proof of concept or prototype. This proof of concept is not in a format that will be ready to release, but this series will help you compose and map a composition and experience it in a playback system. I recommend constantly prototyping your music before moving to the hard coding stage of building a mobile app. This will ensure you have created the best musical experience.

In the Mobile Music Making series, Chapters 10 and 11, we will use *TouchOSC*, *Ableton Live* and *M4L*. You will need to use *Ableton Live Suite* to access *M4L*. You can apply for a 30-day trial of *Ableton Live Suite* at https://www.Ableton Live.com/en/trial/. We use *TouchOSC* to mock up GUIs for prototyping and access sensors built into your smartphone. https://hexler.net/TouchOSC#get

In creating these tutorials, I considered various sensors, such as Arduino and IRCAM RIots[1], as well as other gesture-based sensors. After a thorough investigation and a considerable amount of time, I concluded that *TouchOSC* is the most suitable option for the broadest user base. It is compatible with both PC and Mac computers and any smartphone. Finding a platform everyone can use without incurring substantial costs is quite challenging. *TouchOSC* lets us learn MIDI and OSC messages and access sensors from our smartphones, like the accelerometer and gyroscope. These foundational tutorials aim to provide a starting point in Transmutable Music. While this book opens many doors to specialised areas or new knowledge, the tutorials are specifically designed to familiarise you with working with MIDI, using Open Sound Control (OSC), introducing machine learning algorithms, conducting simple mapping, building a system, and executing complex mapping within your compositional workflow. Ultimately, they aim to facilitate a deeper understanding of the concepts discussed throughout the book.

Tutorial 10.1: Working with *TouchOSC*

TouchOSC Software Requirements

To complete this tutorial, you must purchase a version of *TouchOSC*. The app is currently just over $US12 as of 2024. The app has been developed over several years and is reliable and affordable. I've been using it since 2011, and it has come a long way with fewer bugs. You'll need to ensure you have the latest version of *TouchOSC*, as these tutorials are not compatible with Mark One *TouchOSC*. If you have the latter, you'll need to upgrade. You can find *TouchOSC* on the Apple Store, Google Store and Amazon. Here's a link for your convenience: https://apps.apple.com/app/TouchOSC/id1569996730

https://play.google.com/store/apps/details?id=net.hexler.lex

https://www.amazon.com/dp/B096T636YV

After purchasing the latest *TouchOSC* mobile app, you must download the *Protokol* and *TouchOSC* Bridge software onto your computer. These are both free and can be found here.

TouchOSC Bridge - https://hexler.net/TouchOSC#resources

Protokol - https://hexler.net/Protokol#get

Note – You can download the *TouchOSC* editor for your computer. This will make it much easier for you to design user interfaces. However, this is not required for these tutorials. We will do custom templates and interfaces via the mobile or tablet app.

TouchOSC Overview

The *TouchOSC* app is made up of three main areas that you need to know for this tutorial,

- Toolbar
- Control Surface Editor
- Editor Panel

For an overview of these areas and an introduction to *TouchOSC*, please see Video 1 on the Companion Website or go to the *TouchOSC* Manual here. https://hexler.net/TouchOSC/manual/editor-interface

Tool Bar

The Toolbar is used to create and edit your templates, as shown in Figure 10.1. You can also set up your connections. When using a mobile or tablet version of

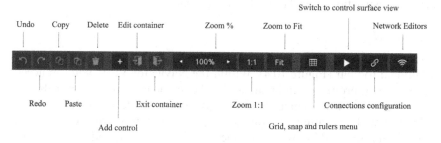

Figure 10.1 TouchOSC Toolbar Overview.

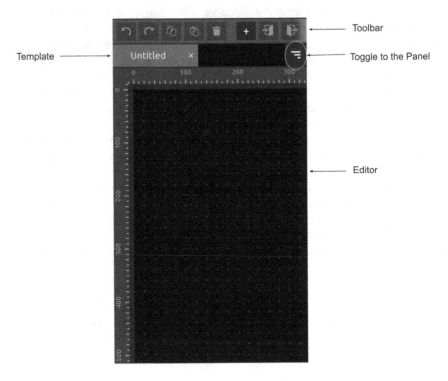

Figure 10.2 TouchOSC Control Surface Editor.

TouchOSC, you must scroll to the right to access the different options on the toolbar.

Control Surface Editor

Here, you can edit your interface or template. To access the Panel area, press the Toggle to the Panel button, circled in white in Figure 10.2.

To add or create a controller, select the + button. It will then appear in the editor. To edit a controller, you need first to select it and then toggle to the Panel area. It is very intuitive to design your GUI or template; for more detailed instructions, see Tutorial 3 in this chapter.

Editor Panel

The Editor Panel consists of the navigator view (Figure 10.3) and an additional toolbar (Figure 10.4). Within the Navigator View, you can access the document properties and a hierarchical document tree of all controls. The Navigator view will change depending on what is selected in the Surface Editor View. For example, if you have added a button to your template and the button is selected in the editor panel view, you can edit and set properties, values, and messages.

Figure 10.3 TouchOSC Editor Panel Screenshots.

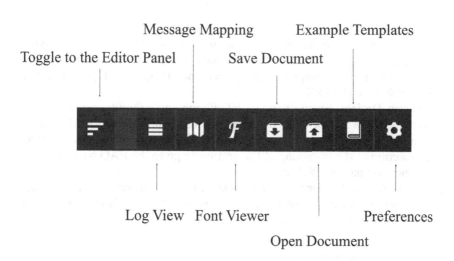

Figure 10.4 TouchOSC Editor Panel Additional Toolbar Overview.

The toolbar in the Editor Panel has the following options, as shown in Figure 10.4. Here, you can save and open templates and access example templates. Press the Toggle to the Editor Panel to return to the surface editor view area.

For details on this, please refer to the *TouchOSC* manual. https://hexler.net/TouchOSC/manual/editor-interface

Configuring TouchOSC and Ableton Live for Mapping

Configuring Your Connection

Your mobile device will send messages over a wireless network to your computer. For this to work, you need to be able to transfer data across your wireless network. If you are using your home WIFI, you shouldn't have a problem. Workplaces and Universities will not allow data to be transferred across their wireless network for security reasons. Potential hackers, for example. If you work from a university or workplace, use a hotspot (a router not connected to the internet) or USB cable to transfer this data (Granieri, 2020).

You can also use your Mobile Hotspot; however, you will need a third device. This is a good option if you have a second mobile or tablet with *TouchOSC* installed. Unfortunately, you can't use your mobile as a hotspot to transfer data from *TouchOSC* (Granieri, 2020).

For more details on how to set up your devices for *TouchOSC* if using a hotspot or USB cable, please see the FAQ on *TouchOSC* on the Companion Website.

Setting Up

1 **Launch the applications**: Start '*Protokol*' and '*TouchOSC* Bridge' on your computer. *TouchOSC* bridge does not have a window GUI; a B icon on the top right of your computer will appear once it opens – Figure 10.5 is a screenshot of *Protokol*.
2 **Connect Devices**: Ensure your mobile phone and computer are connected to the same WiFi network. If you cannot use your WIFI network because you are at a university or workplace, see the section on configuring your connection at the beginning of this section or go to the FAQ on *TouchOSC* on the Companion Website.
3 **Access *TouchOSC***: Open the *TouchOSC* app on your phone. For a guided tour, watch 'Video 1' for essential navigation tips or check out the app overview earlier in this chapter.

Configuration Using a Wireless Network

4 **Bridge Connection**: Tap the connections configuration button on the toolbar, which looks like a 'link' icon in *TouchOSC*; see Figure 10.1. If

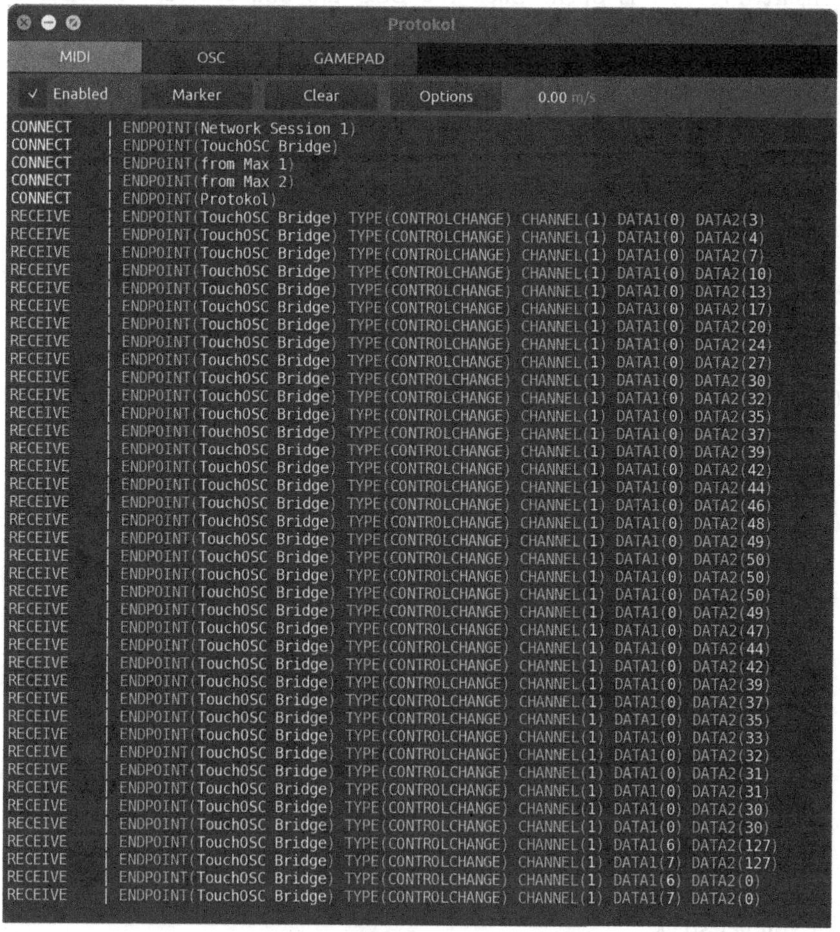

Figure 10.5 Screenshot of *Protokol* – MIDI Tab.

you cannot see the link button, use your finger to scroll the toolbar horizontally to the right until you reach the end.

5 First, connect the app to the bridge. Select Bridge, click Browse, and select your computer. The Host field will now list your computer's IP address. If you can find your computer in the list, make sure TouchOSC Bridge is open on your computer.

6 Click on the back option in the top left of your screen to return to the Connections Configuration area.

7 Next, we will set up the OSC connection. Click on the OSC menu option. Make sure the Connection 1 box is ticked. Tap the Browse button and select your computer and port number. If you can't see your computer,

ensure *Protokol* is open. Click the OSC tab and ensure the tick box Enabled is on. The port number will be 8000 – no need to change this now.

8 Finally, we will set up the MIDI connection. Click back to the connections Configuration area and select the MIDI option. Ensure the Connection 1 box is ticked and select browse next to the send port. Select <Bridge 1>. There is no need to set up the Receive Port at this stage. However, you can select <Bridge 1> for this too. This is important if you send data back from your computer to *TouchOSC*, which we won't be covering as part of this tutorial series.

Open a Template on TouchOSC

9 **Access the Editor Panel**: In *TouchOSC*, tap the 'toggle to panel' button to see the different control options. See Figure 10.2 for the correct button.

10 **Choose Template**: Tap the example templates button in the additional panel toolbar. It has a book icon, see Figure 10.4. Select 'Simple Mk2' as your control surface template.

11 **Start Session**: Press the 'play' button in the app's top toolbar to initiate the GUI. Explore the template by interacting with the controllers. Four 'pagers' (or tabs) exist: Faders, Pads XY and Matrix.

12 **See the MIDI messages**: Go to the *Protokol* application and click the MIDI tab. As you interact with the *TouchOSC* template, notice the MIDI messages sent to *Protokol* as per Figure 10.5.

13 **See the OSC messages**: Next, click on the OSC tab in *Protokol* and observe the OSC messages sent to *Protokol* when interacting with the *TouchOSC* Simple Mk2 template.

Ableton *Live Setup*

8 **Load Project**: Open the *Ableton Live* Set from the Companion Website, found in the 10.2 Tutorial. Note that the completed set is just for reference if you get stuck.

9 **Configure** MIDI: In *Ableton Live*, go to preferences (use Command then ',' on Mac or Control then ',' on PC), select 'MIDI', and ensure *TouchOSC Bridge* is active for both input and output, as shown in Figure 10.6. This

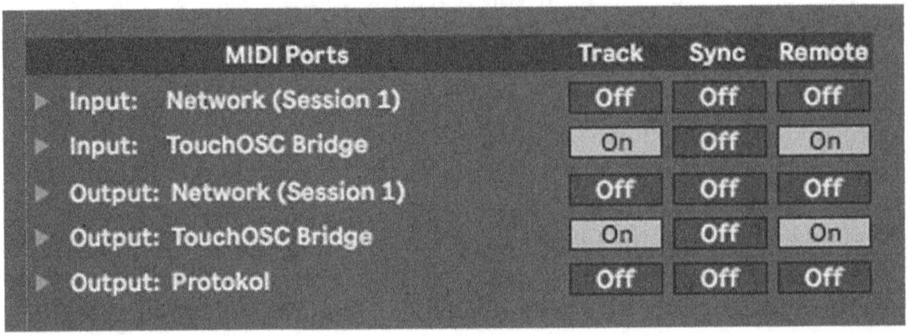

MIDI Ports		Track	Sync	Remote
▶ Input: Network (Session 1)		Off	Off	Off
▶ Input: TouchOSC Bridge		On	Off	On
▶ Output: Network (Session 1)		Off	Off	Off
▶ Output: TouchOSC Bridge		On	Off	On
▶ Output: Protokol		Off	Off	Off

Figure 10.6 Screenshot of *Ableton Live* MIDI Preferences.

will allow you to receive and send MIDI messages via *TouchOSC Bridge*. We will look at how to receive OSC messages using Max and Max4Live in Tutorial 4.

The first tutorial was a short introduction to the functionality of *TouchOSC* and how to connect your computer to send OSC and MIDI messages over a wireless network. The following tutorial will review how to map MIDI messages in *Ableton Live* to control any parameter.

Tutorial 10.2: Simple Mapping to Musical Parameters (MIDI)

Exercise 1 – Mapping *MIDI* in Ableton Live

1 Ensure you have set up *TouchOSC* and *Ableton Live* per the Tutorial 1 instructions. Also, ensure the *TouchOSC* bridge and *Protokol* applications are open on your computer.
2 Ensure your Simple MK2 template is in play mode and not edit more. Do this by tapping the play button on the top toolbar.
3 Open the Ableton Live Set on the Companion Website if you haven't already.
4 Make sure your MIDI preferences are set up like Figure 10.6. The *TouchOSC* Bridge is acting like a MIDI controller within Live. Ensure the Track and Remote options are On (yellow).
5 Now, we are ready to start. First, we will map the Faders tab of the Simple MK1 template to the session window of your *Ableton Live* set.
6 In *Ableton Live*, click on the MIDI mapping button on the top right of the transport bar or as a shortcut, press command and 'm' (control 'm' for PC). Your *Ableton Live* set will now have a blue tint to all the controls that can be mapped and controlled.
7 In *Ableton Live*, mouse click on the gain (volume) slider of the Synth 1 track. Now move Slider 1 on the *TouchOSC* Fader page. The slider is directly mapped on the right MIDI mapping panel. See Figure 10.7.
8 Click MIDI mapping off again and test that the gain on track one can be controlled using *TouchOSC*. Press command and 'm' (control 'm' for PC) for a shortcut.
9 Now map all four tracks' gain sliders to the four sliders in *TouchOSC*.
10 Next, we will map the gain in the master track in *Ableton Live* to the top horizontal slider in *TouchOSC*.
11 Finally, we will map the play and stop buttons in *Ableton Live* to the green toggle buttons beneath the sliders in *TouchOSC*.
12 Now, turn off MIDI mapping in *Ableton Live* and use the app to control your mapped parameters.

Use the following questions to evaluate the experience we have mapped in exercise 1.

Doing this at the end of each exercise in this tutorial would be best.

1 With this functionality, how would you rate the user experience?

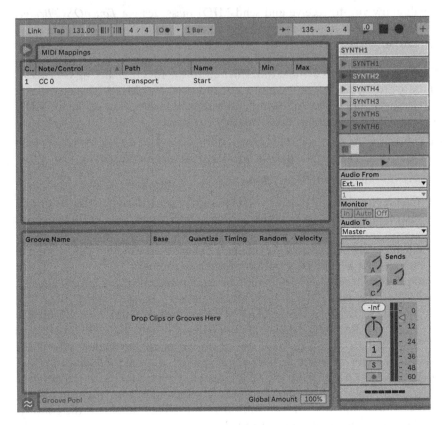

Figure 10.7 Screenshot of MIDI Mapping Example in *Ableton Live.*

2 How long would someone want to engage with this idea, and how could it be improved?
3 Why isn't this an interesting example of an interactive work?
4 How different can the track sound?

In the next exercise, we will try different mapping options at the macro level (using stems) to create more variability or musical options for the user to experience.

Exercise 2 – Extending the Structure (One-to-One Mapping)

Now, we will add the ability to change song sections to see if we can create more options.

1 On the Simple Mk2 template, click on the 'Pads' page (tab) on *TouchOSC*
2 The 16 pads do not toggle on and off like the green buttons on the 'Faders' Page.
3 Go to Protocol to the MIDI tab, press one of the buttons on the Pads page, and notice it sends two messages. What this data tells us is that the message

is sent to *TouchOSC* Bridge, the type of MIDI message is a note, MIDI Channel 1, Note 36, Velocity 127, and the second message is the same except for the DATA2 or velocity value, which is 0. This is equivalent to sending a NoteOFF message.

RECEIVE|ENDPOINT(*TouchOSC* Bridge) TYPE(NOTEON) CHANNEL (1) DATA (36) DATA2 (127)

RECEIVE | ENDPOINT(*TouchOSC* Bridge) TYPE(NOTEON) CHANNEL (1) DATA (36) DATA2 (0)

4 Now click on the Faders tab and click on one of the green buttons. Notice that we are sending a Control Change MIDI message this time, not a Note Value. Also, we need to press the button again to turn the button off.

RECEIVE | ENDPOINT(*TouchOSC* Bridge) TYPE(CONTROLCHANGE) CHANNEL (1) DATA (5) DATA2 (127)

RECEIVE | ENDPOINT(*TouchOSC* Bridge) TYPE(CONTROLCHANGE) CHANNEL (1) DATA (5) DATA2 (0)

It is essential to understand that in *TouchOSC*, we can have a button called 'Momentary' or 'Toggle'. We can also choose to send a note or control change MIDI command. We will look at these options in more detail in the following tutorial when we design our own templates in *TouchOSC*.

5 Now go back to *Ableton Live* and change to the Arrange window (Press tab). Notice that a version of the track has been recorded into the arrange window. Also, I have set up locators at different section changes.
6 We will map the pads on the Pad tab of *TouchOSC* to each of the locators in the arranged view of the *Ableton Live* set. To do this, use MIDI Mapping in *Ableton Live*, Command and 'm' (Control 'm' for PC). Click on the locator and then select one of the pads. Make sure you only map one locator per pad, as per Figure 10.8.
7 In Figure 10.8, you can see the mappings made in *Ableton Live*. Notice again the different MIDI commands. The first slider mapping to the Track 1 Gain is a control change. The pad we just mapped sends notes; for example, A0 is mapped to the A4 locator.
8 On the transport section of *Ableton Live* is the global launch quantisation; at the moment, it is set to 1 bar. This means it won't happen instantly when we trigger a sample or locator; it will wait until the end of 1 bar. This is great for keeping the music sounding good and on time. However, when designing a user-based experience, you want to ensure that the change is instant when they tap the pad. You can do this by changing Global trigger settings to ½ note and experimenting with the correct response time.
9 Explore this extra functionality and play with the new mappings. Try to imagine this was an app that you were playing with. Does it improve the experience from exercise 1? Does the track sound good still?

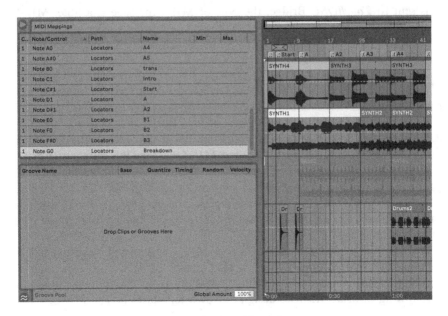

Figure 10.8 Screenshot MIDI Mapping in *Ableton Live*.

Exercise 3 – *MIDI* notes (One to One Mapping)

1 Now, we will use pads to control a MIDI instrument in *Ableton Live*. Delete the previous locator mappings you have made. To delete a mapping in *Ableton Live*, go into MIDI mapping in the *Ableton Live* project and click on the mapping you want to delete in the MIDI mappings area. Then press delete. See Figure 10.8.
2 You can keep playing the audio from the arrange window or return to the session view and select some loops. In this example, we are just playing MIDI notes.
3 Find the Wave Evolve Pad Wavetable instrument in the browser and double-click on it to create a new armed MIDI track.
4 Play notes on using the Pads.
5 Now try with a drum kit. Arm a new MIDI track with a drum kit assigned. I am using the Full Lotus Kit. Using the top pads, trigger the instrument. Note that the pads in *TouchOSC* are sending MIDI to C0, and no samples are attached to those notes in the Full Lotus Kit instrument. In the following tutorial, we will look at how to change the note values sent from *TouchOSC*.

Exercise 4 – *Adding More Parameters (One to Many Mappings)*

We will now create effect racks for the Drums, Synths and Bass to allow more variability in the rendered musical output.

Drum Parameter Options

(a)

(b)

Figure 10.9 Screenshot of Mapped Drum FX Audio Effects Rack.

1 Add an empty Audio Effect rack device in *Ableton Live* on the drum track. You can do this by going to Audio Effects in the Browser and selecting Audio Effects Rack. You could also search for Audio Effects Rack in the browser.
2 Rename the device Drum FX. See Figure 10.9.
3 Add an auto filter to the device, and explore changes you can make.
4 Next, add a 16th note delay, explore some options or use the following – left on one and right on 4.
5 Add convolution reverb; this device is in the Max4Live audio folder in the browser; explore 'type' options or choose 'Swedish Nuclear Reactor' – see Figure 10.9.
6 Open the Macro section of the Audio Effects Rack, which is the left icon of the device in the grey area). Map the 'dry/wet' dial of the delay device to macro 1. (By right mouse click on the 'dry/wet' dial) – Figure 10.9.
7 Next, map the Reverb device's 'dry/wet' dial to macro 2, and finally, the auto filter's frequency to macro3 and the resonance to macro 4. In this exercise, we use mapping within Ableton Live to create variability. See Figure 10.9.

Synth Parameter Options

1 Add a Convolution reverb to the Synth2 track.
2 Next, add an 8th note delay device to the group
3 Finally, add an auto filter device to the track.
4 Play around with the settings with the Synth2 track on solo. Explore how you can change the sound of the synth using these devices.

(a)

(b)

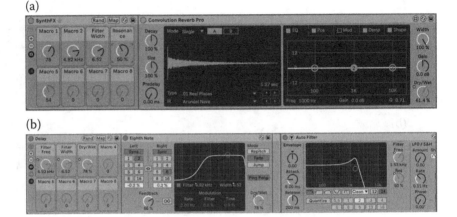

Figure 10.10 Screenshot SynthFX Audio Effects Rack Mapping.

5 Now, we will map some of these parameters. Select all the devices by holding down 'shift,' then right-clicking the mouse and selecting the group. This is another way to create an audio effects rack. The quick way to group is command and 'g' (control 'g' on a PC).

6 Rename the device 'SynthFX'.

7 As we did in the Drum effects rack, we will map parameters to the Macro area of the audio effects rack. However, we will map two parameters onto the same dial this time.

8 Right-click on the 'dry/wet' dial of the Convolution reverb device and select 'Map to Macro 1'; now, right-click on the filter parameter of the auto filter device and select 'Map to dry/wet' (which is the macro 1). Move the Macro 1 dial of the Auto filter and observe that you are controlling two parameters. See Figure 10.10.

9 Next, map the filter in the delay device to Map to Macro 2 and the Width to the Map to Macro 3. See Figure 10.10.

10 Finally, Map the Delay's 'dry/wet' dial to 'Map to Filter Freq' and map the 'res' in the Auto Filter to 'Map to Macro 4.' See Figure 10.10.

11 Use the other *TouchOSC* controls on the SimpleMk2 template and experiment with mappings. Make sure you try the XY pad and the Matrix. Note that although you can map more than one parameter to macro dials in *Ableton Live*, you can't map a *TouchOSC* MIDI message to multiple parameters in *Ableton Live*. This will confuse *Ableton Live*, and the template won't work well. Ensure that any controller in *TouchOSC* is only mapped to one parameter in *Ableton Live*, like a slider, dial, or button. You can, however, map a MIDI message from *TouchOSC* to a dial in *Ableton Live* that controls multiple parameters.

12 Once you have mapped these extra variability options, play with the *TouchOSC* template and try to make a track. During this time, think about the user experience. How could you improve this? Do these extra parameters add a more diverse rendered output?

Exercise 5 – One to Many Mappings (Self-Directed)

Sometimes, you can map a controller to multiple devices. For example, you may want to assign an X-axis with various parameters like a filter and wet/dry parameter. You can extend your options using the following Max4Live patches.

XY Pad M4L[2]

You can map one control parameter to an X and Y pad. For example, the X could be mapped to a filter and the Y to the resonance.

MultiMap M4L[3]

This is an excellent patch to map multiple parameters to one controller. You can map your X-axis to various inputs. You must add another MultiMap to assign different parameters and control your Y-axis.

1 Firstly, explore some of the other templates in *TouchOSC*. You can find these by entering the editor panel and selecting the book icon.
2 Try experimenting with mapping options with this *Ableton Live* set and other templates.
3 Also, try mapping 'one to many' using the Max4Live mapping devices introduced in this exercise.
4 You could also map the pads to the clips in the session view to trigger new clips.
5 Consider controlling this *Ableton Live* set with a *TouchOSC* template to consolidate mapping approaches in this tutorial.

In the following Tutorial, we will look at designing your own simple *TouchOSC* templates. This will help you to create prototypes for user-based audience experiences.

Tutorial 10.3: Designing a Template/GUI with *TouchOSC* (MIDI)

This tutorial is presented as a series of videos, which can be accessed on the Companion Website. It covers how to use *TouchOSC* to design your own templates. We cover user experience techniques and controlling mapping with MIDI messages in more detail.

Notes

1 https://ismm.ircam.fr/riot/
2 https://maxforlive.com/library/device/3580/xy-pad-fine
3 https://maxforlive.com/library/device/4502/multimap-pro

Reference

Granieri, N. 2020, January 7. *Effective OSC Communication between macOS and Mobile.* Medium.

11 Mobile Music Making Advanced Tutorials (OSC)

Tutorial 11.1 – Accessing Accelerometer Data in Max4Live (OSC) – Working with LOM

This tutorial will explore using sensors and OSC (Open Sound Control) to capture and utilise accelerometer data. This measures movement and orientation. Specifically, we will focus on receiving this data in a software called *Max*. Once we have the data in *Max*, we'll convert it into a form that can be used to change or 'modulate' sounds within a *Max* for *Live* (*M4L*) device in *Ableton Live*.

You can find additional help and walkthroughs in support videos on the Companion Website for all the tutorials and exercises in this chapter. It's important to note that the upcoming tutorials require you to have some basic programming knowledge in *Max*. If you are new to *Max* or need more confidence in your *Max* programming skills, you might find learning easier by watching the video tutorials on the website instead of reading the text explanation here.

Mobile Phone Sensors

Just about everyone carries a smartphone with them. Smartphones have a variety of sensors that we can access. These include accelerometers, gyroscopes, magnetometers and attitude sensors. *GyroOSC*, available for iPhone, can also access the compass, rotation matrix, rotation rate, quaternion, gravity, magnetic field, GPS, and altitude.

As well as sensors, your smartphone can send GPS coordinates and access APIs on the internet, such as NASA space information, weather data, and maps.

Accelerometers

Accelerometers are sensors that tell how quickly your phone moves or changes speed. They check this speed change along three different lines, called the x, y, and z axes. You can think of these as imaginary lines that run through your phone in different directions. For example, the accelerometer can feel the movement when you are on a train speeding up. If you turn around inside the train to face the opposite direction, the phone's

DOI: 10.4324/9781003273554-15

accelerometer will sense that the direction of the speed change has also turned around, just like you did.

Exercise 1 – Receive OSC Messages in Max4Live

In this tutorial, you will learn how to:

- receive OSC messages in *Max* and *M4L*
- extract the data you want to use from an OSC message
- scale the data to work with a dial using MIDI.

Help Resources: If you get stuck, check out the how-to videos on the Companion Website for visual assistance.

Setting Up: We are using port 9000 to send and receive OSC messages. You can keep this port or change it if you need to. To change it:

- Go to *Protokol*, find the OSC tab, and enter a new port number.
- Update the same port number in your *TouchOSC* app settings to connect.

Using *Max*: *Max* uses **UDPreceive** and **UDPsend** to handle OSC messages. (Refer to Chapter 6 Pg for a basic understanding of OSC).

1 Open *TouchOSC* on your mobile and launch *TouchOSC* Bridge, *Protokol,* and *Ableton Live* Live on your computer.
2 Open the simple Mk2 template on *TouchOSC*.
3 Ensure you receive data in *Protokol* in the MIDI and OSC tabs.
4 Close *Protokol* or you can uncheck the enabled option in the OSC tab.
5 Open a new live project and create a new *M4L* audio effect
6 Delete everything inside of the patch.
7 Press 'n' to create a new object.
8 Type 'udpreceive 9000' into the new object.
9 Now, create a new object and type print.
10 Connect the 'udpreceive' object to the print object and open the *Max* console. See Figure 11.1
11 Use the 'Simple Mk2' template in *TouchOSC* to send OSC data to *Max*. See the Figure 11.1. If you are not receiving data, go to the trouble-shooting section at the end of this exercise.

Understanding Data: The data (OSC message) you receive looks like **/1/fader1 0.808734**, indicating the position of a fader control.

Extracting Data

12 Use a 'route' object with an argument of the part of the message we want to remove.Type route /1/fader1.

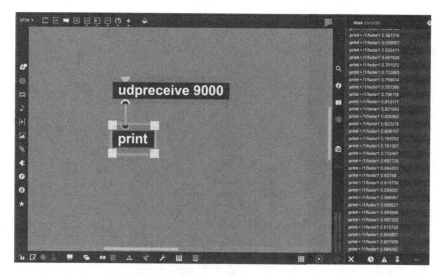

Figure 11.1 Screenshot of *M4L* Patch.

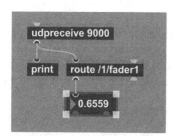

Figure 11.2 Screenshot of *M4L* Patch

13 Create a new message by pressing 'm' and connecting the route object's left outlet to the **right** inlet of the message.

14 Connect the left outlet from the 'udpreceive' object to the left inlet of the route object. As shown in Figure 11.2.

15 Now move fader 1 on *TouchOSC*. We will see the float value.

16 We can collect and look for values from multiple OSC messages in the route object. Add the following arguments to the route object /1/fader2 /1/fader3 /1/fader4. Each argument should be separated with a space. As shown in Figure 11.3.

17 The right outlet of the 'route' object will send any messages that do not have the addresses we are looking for. See Figure 11.3.

Connecting to a Dial and Scaling Data

18 Now we can use the fader in *TouchOSC* to control a dial in *Max*. Add a new 'dial' object.

19 Attach the fader1 data to the 'dial' object and try to control it with your TouchOSC fader 1. It is not working because the dial is set up to use

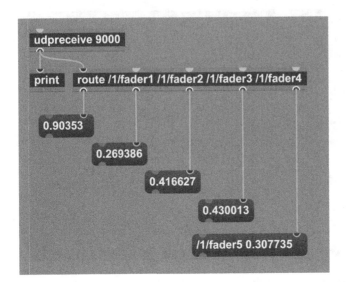

Figure 11.3 Screenshot of *M4L* 7 Patch.

MIDI data. This has a range of 0 127. However, this fader has a range of 0 1. You can see the values of an object in the inspector window.

20 We can change the scale of the data from 0 1 to 0 127 using the scale object. Add a scale object with the arguments 0 1 0 127. See Figure 11.4.

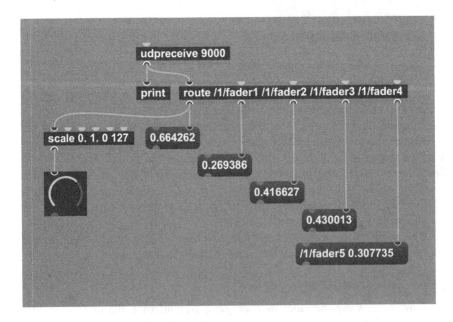

Figure 11.4 Screenshot of *M4L* Patch.

21 We can now control the dial with our *TouchOSC* fader. We have also converted OSC to MIDI.

22 Save your *M4L* device and arrange the objects neatly for easy viewing. Close *Ableton Live* and *Max* when done.

Troubleshooting – When Data Is Not Received in Max

Important – when using *Protokol* and *Max* software simultaneously, interference with each other might occur because they use the same communication port. If you notice that you are not getting any OSC (Open Sound Control) data, it could be because the port is getting mixed signals from both programmes. To fix this, try closing both *Max* and *Protokol*, then reopen just one. This should clear up any confusion and allow you to receive OSC data properly.

If you are having trouble with the *Protokol* and *Max* applications receiving data, another solution is to change the port number you are using. Here's how to do it:

In the *TouchOSC* app, tap the 'link' icon to your OSC settings. This is where you can change the port number for sending data. If you need to receive data in *Max*, you will update the port number in the 'UDP Receive' object within the *Max* programme. Similarly, if you are using *Protokol*, you should also update the port number there.

It's a good practice to choose a port number higher than 8000 because many of the lower numbers might already be used by other applications on your computer. Using a higher port number reduces the chance of conflicts with other programmes and ensures that your data is sent and received correctly.

Exercise 2 – Build a New **TouchOSC** *Template That Can Send Accelerometer Data as OSC Messages*

In this exercise, we will create a new *TouchOSC* template that can send accelerometer data from your phone as an OSC message. Remember that Video Tutorials are available on the Companion Website if you need help.

1 Create a new *TouchOSC* template. In the editor panel, press the open document; see Figure 10.4. Select New at the bottom of the screen.

2 Close the open document page in the editor panel area, or you will not see the Navigator section.

3 In the Navigator section of the editor Panels area, scroll down to Script and expand the script area. *TouchOSC* enables you to develop its design and interface options using *TouchOSC*'s scripting API, which is based on Lua 5.1[1]

4 Add the following code exactly.

```
if(hasAccelerometer()) then
update = function()
local values = getAccelerometer()
sendOSC('/accxyz,' table.unpack(values))
end
end
```

5 Press the play button on the bottom right of the screen. This will action the code that has been entered. It will also highlight any errors in red. The code will not work with any errors. Check and re-edit the code to make sure it is correct. Make sure you press enter at the end of each line as above.

6 Next, you want to play your new template using the play button in the top toolbar. You may need to scroll to the right to find it.

7 Go to *Protokol* and make sure you are receiving the accelerometer data. It will be continuous data. You may need to open *Protokol* if you closed it in the last exercise. Make sure *TouchOSC* is connected to the correct port, and *Protokol* is receiving data. If not, go to the link icon in the OSC menu option and click browse to select your computer's IP and correct port.

8 Save the *TouchOSC* template 'Tutorial4.' You can save a template by exiting the play mode using the top circle on the right and then pressing the toggle to the editor panel. Tap the save document button on the fourth from the right. Enter the new name at the bottom of the screen where untitled is written. Then tap the tick button to the right.

Exercise 3 – Organising the Data

1 Set up *TouchOSC* so that *Protokol* is receiving OSC accelerometer data.

2 Go to *Protokol*, OSC tab and click off the enabled tick to stop *Protokol*, accessing port 9000. The accelerometer data will stop. Note – you must have OSC enabled on *Protokol* to connect your phone to the network/ computer.

3 Open a new *Ableton Live* set.

4 Add an empty *M4L* audio effect to the first MIDI track.

5 Open the patch in *Max*.

6 Press the presentation icon to go into the patch and ensure it is in edit mode.

7 Delete everything in the patch, as we will not be using it.

8 Add a new object by pressing 'n.'

9 Type 'udpreceive 9000.'

10 Connect a message to check that we are receiving the OSC data. See Figure 11.5.

11 Add a new route object to remove the /accxyz text. See Figure 11.5.

12 Now let's unpack this data to access the x, y, and z float separately. Create a new unpack object with the arguments f f f.

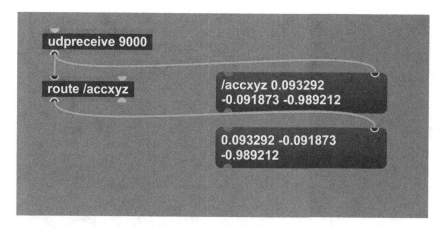

Figure 11.5 Screenshot of *M4L* Patch.

13 Connect three new float objects (press 'f' for a shortcut) to see the individual data set.

14 We will work with the x value. Add a 'live.slider' object to control with our x value.

15 Note that it is not working as the accelerometer's range is approximately from -6. 6. (Different smartphones will have different accelerometer sensors. You will need to check the range of your accelerometer. You can probably google this.)

16 Let's adjust the slider's range to '-6. 6.' in the inspector instead of '0. 1.'

17 Is it functioning well? Work out the direction you need to move your phone to make the slider work on the x-axis. By looking at the behaviour of the data while doing a specific action, we can understand the effect our action has on the data. However, it is not easy to do this because of the constant data being received.

18 Next, change the range of the 'live.slider' to '0. 1.' in the object's inspector. See if you can use your phone to control the fader more easily.

19 You may notice you can move the phone in other directions on the x value without moving the slider. Now add another 'live.slider' object with a range of -1. 0. We have two actions using the x data. See Figure 11.6.

20 Now, we will update the presentation of our *M4L* patch. Select both sliders, right-click, and select Add to Presentation.

21 Click on the presentation icon. It will turn yellow. Move the sliders to the top left. Save.

22 Close the patch and go back to *Ableton Live*.

23 You need to set up the patch to display in presentation mode when it is a device in *Ableton Live*. To do this, open the patch in *Max*. Right-click anywhere on the patch and select Inspector Window.

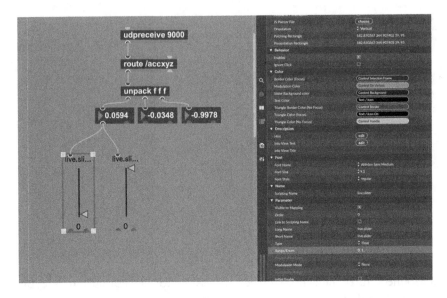

Figure 11.6 Screenshot of *M4L* Patch.

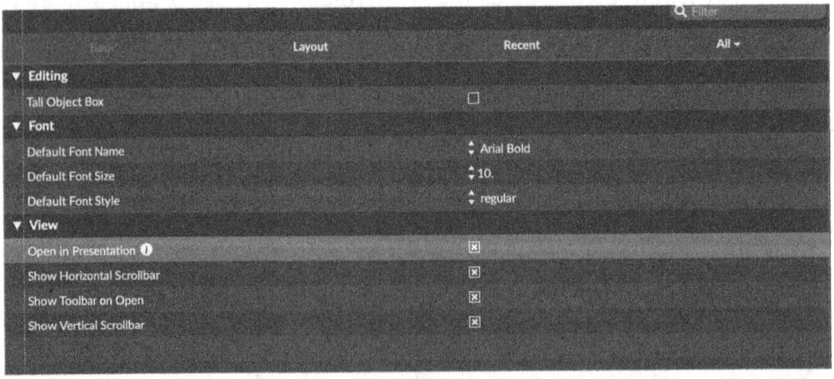

Figure 11.7 Screenshot of *M4L* Patch.

24 In the Inspector Window, click on the box in the Basic tab under View to arm the Open in Presentation option. See Figure 11.7.
25 Save it. You will now see your patch with the two sliders in the Device view of *Ableton Live*. See Figure 11.8.
26 Now try adding more sliders on the x, y, and z values and create more gestures.

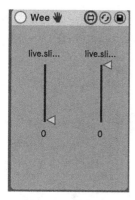

Figure 11.8 Screenshot of *M4L* Patch.

Exercise 4 – Use Your Phone's Accelerometer Data to Control Sound

To save time in this Exercise, we will map the data directly to parameters within *Ableton Live*. We can do this with a quick hack using the APIs within the *M4L* Building Tools pack.[2] You can download this pack from the *Ableton Live* website's packs section. It is free. The pack contains 138 *Max* devices and tutorials. Make sure you explore these if you haven't used this pack before.

1 Open a *Max* Api 'Amap4' from *M4L* audio effects or the *M4L* Building Tools pack. This device can map the control dial to 4 different parameters and provide extra data controls. For example, smooth, jitter, inverse, and range. It contains a bpatcher.[3]
2 Open the patch in *Max* and look at how it has been created. Click the presentation view off to go to the edit area of the patch. This patch has been frozen. Click on the snowflake icon. This will be blue to unfreeze the device.
3 *To open the Bpatcher interface, right-click on 'object' and then 'OpenOriginal M4L.SignalToLiveParam.Maxpat' This will open the original Bpatcher patch.* Make sure you close this before moving to the next step.
4 We will use our accelerometer data on the Z-axis to control the dial in this patch. But first, we must save our hacked device, 'z valueTutorial5.'
5 After we have saved a new device, we will need to create the same objects from Exercise 2.
6 Create and connect the following objects: udpreceive 9000, route /accxyz and unpack 0. 0. 0. (The 0. can also be used for a float value). See Figure 11.9.
7 Add a float object (f) and attach it to the z output. See Figure 11.9.
8 Connect the outlet of the float to the inlet of the control dial. See Figure 11.9.

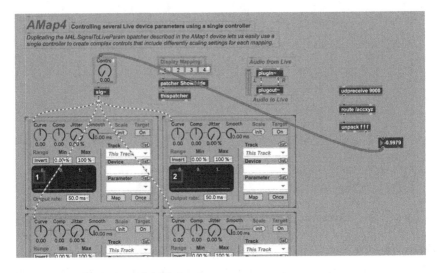

Figure 11.9 Screenshot of *M4L* Patch.

Mapping to Operator Instrument in Ableton Live

8 Now save the device and close it so it opens in *Ableton Live.*

9 Next, we will map this to the *Ableton Live* operator device. Add a new empty operator instrument device to the **second** midi track. Note – the *M4L* device will need to be on its own track.

10 Map the first 1st display mappings to the **Osc1 level** in the operator. To do this, click on Display Mapping 1 so it is yellow. Then click on the grey 'map' button on the bottom left of the device. Now, you can select the parameter you want to map to.

11 Select the 2nd Display mapping on the device and map it to the Filter Frequency in Operation. and

12 Finally, map the 3rd display mapping to the 'Osc2 level' in Operator.

13 Move your phone on the z-axis and see your mapping dials move. However, we still have no sound. Let us make a midi note in *TouchOSC.*

Creating More Functionality in Your TouchOSC *Template*

14 Create a new button on your 'Tutorial4' template. Tap the '+' icon in the toolbar in the surface editor area, and select 'button.' For help with this, see Tutorial 11.1 on the Companion Website. Set up the button to send a MIDI note. (You can increase the size of this button; see Figure 11.12.)

15 Select the 'button' and tap the 'toggle to editor panel.'

16 In the Navigator area, expand 'Messages' and 'MIDI.' Change the 'type' to Note_On. Then, change 'Index' to constant in the' Note' field. Finally, type the note value of '72.' See Figure 11.10.

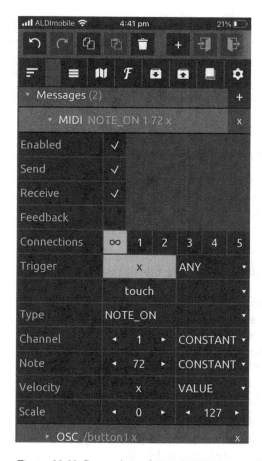

Figure 11.10 Screenshot of *TouchOSC* 'Button' Editor Panel Area.

17 Add a limiter (Audio effects/Dynamics) to your master track to protect your ears.
18 Make sure the MIDI track with the operator instrument is armed. (Record button on.)
19 Try holding the midi button and moving your phone along the z-axis.

Expand the Sound Design with Another MIDI Instrument

20 Add another button to your *TouchOSC* template with a note value of 60. Same as we just did.
21 Have a play with the sounds you can make.
22 Add the 'descending dreams' wavetable instrument to a new MIDI track.
23 Map the 4th display mapping of your *M4L* device to the Osc 1 position on the 'descending dreams' wavetable instrument. See Figure 11.11.

Oscillator 1 Position

Figure 11.11 Screenshot *Ableton Live* Wavetable Device.

24 Arm the record icon on both MIDI tracks (operator and wavetable) using the command and clicking the record button on each track.
25 Explore the sounds you can create now.

Create Another Device for the X-axis

26 Let us add some more functionality using the x-axis now.
27 Duplicate your 'zvalueTutorial4' *M4L* device.
28 Open it and save it as x valueTutorial4.
29 Use the x value data to connect to the control dial instead of the z value and save. See Figure 11.8. Instead of connecting the last value from the unpack object. Connect the first float value and connect it to the control dial.
30 Close your *M4L* patch and go back to *Ableton Live*.
31 Test how to control your new device by moving your phone on the x-axis.
32 Now map your new x valueweek4 device to the following:

   1 map – Operator – Osc3 Level.
   2 map – Descending Dreams – Filter 1.
   3 map – Operator – LFO rate (turn on LFO).

33 Have a play with the types of sounds you can make.

Extend the Sound Design with Loops and Buttons

34 Add a drum loop in clip 1 Audio track 3 on audio 1 (Drum groove HalfHeart 116).
35 Add a bass loop in clip 1 Audio track 4 (Bass Bounce C 128bpm).
36 Create four new buttons in your *TouchOSC* template. There is no need to change the message the buttons send out, as we will MIDI map in *Ableton Live*. See Figure 11.12.
37 Like in Tutorial 2, use MIDI mapping (Command 'm' or Control 'm' on a PC). Map two new buttons to the clip trigger triangle of the two loops. Then, map the other two buttons to the clip stop button of the corresponding track.
38 You can change the colours of the buttons if you like.
39 Now, you can trigger loops to add to your mobile sound experiment.

Figure 11.12 Screenshot of *TouchOSC* Template

40 Create some more control, sound, and mapping based on what we have learned in this tutorial. For example, add more functionality with the XYZ data, as in Exercise 3, and make more control dials and Bpatchers.

Tutorial 11.2 – Developing Simple Models with Decision Trees
Max4Live (OSC)

What You'll Learn: In this tutorial, we will use the accelerometer in your mobile device to create a drum machine that you can control by moving your device. You'll learn how to set up random and predictable sound variations, giving you creative ways to make music. We'll simplify the accelerometer data to make it more useful for our purposes and dive into how *Ableton Live* can be controlled with *Max* for *Live* (*M4L*) using the Live Object Model (LOM).

What is LOM? The Live Object Model[4] is like a map of *Ableton Live* that allows *M4L* to access and control its different parts.

Skills Gained: You'll acquire skills that are not just limited to drum machines. They can also be applied to other sensors and types of data.

User Experience: While this tutorial, similar to the previous one, may not give you a polished user experience, it is valuable for teaching methods to manage and use data musically. The following tutorial, Tutorial 6, will address these user experience issues by introducing machine learning to create a smoother interaction.

Where to Find More: For the entire series and additional resources, visit the Companion Website.

Tutorial 11.3 – Intro to Machine Learning Using Accelerometer Data in *Max4Live*

What You'll Discover: In this tutorial, you will get a primer on machine learning as it applies to music creation using a programme called Wekinator. Drawing inspiration from Rebecca Fiebrink's online course,[5] you will explore ways to apply different machine learning algorithms for handling both discrete and continuous data. This knowledge will be explicitly linked to using accelerometer data from your mobile device through *TouchOSC*, *M4L*, and *Ableton Live* Live. You'll learn to send OSC data and craft more sophisticated *M4L* devices for an improved user interaction experience.

Skills Gained: You'll gain an understanding of:

• integrating machine learning into your music projects.
• sending and interpreting OSC data for creative purposes.
• designing *M4L* devices that respond intelligently to user interactions.

Where to Find More: The complete series of tutorials, extra materials, and helpful links are available on the Companion Website.

Creative Exercise: Crafting Your Interactive Music Prototype with Mobile Integration

Objective: create your own interactive music system prototype that leverages the unique capabilities of your mobile device. This exercise invites you to combine the convenience of *TouchOSC* with the power of *Max* for Live and *Ableton Live* Live to construct a user-interactive musical experience.

The Challenge: Design and compose an interactive prototype that embraces user participation. Your system should incorporate sound synthesis, sampling, and programming elements to forge an interactive prototype that thrives on variability, ensuring a unique auditory experience with each playback. The extent of user control within the system remains at your discretion.

Technical Composition: Develop a custom *TouchOSC* template that acts as your interactive system's control centre. Your musical arrangement should be dynamic, using a blend of OSC data (leveraging the accelerometer and gyroscope) and MIDI data. Although integrating Wekinator is optional, it can enhance your project's interactivity.

Tutorial Guidance: The content from the tutorials in Chapters 10 and 11 will be instrumental in guiding you through the construction of this project. Complement this with your research to deepen the development of your work. The musical content should be neatly organised within an *Ableton Live* Live set, featuring a rich mix of audio, MIDI, DSP, and synthesis devices. Create custom *Max* for Live patches to map OSC data within Live and perform intricate audio or MIDI tasks.

Design and Evaluation Essentials

Composition: Evaluate the piece's variability and effectiveness in the intended form. Assess the production quality of loops, sound design, effects, layering, and the organisation and mapping of data to musical parameters. Confirm that the *Ableton Live* set and *Max* for Live patches are meticulously organised and designed.

User Interface Design (UI) & User Experience (UX): Ensure your prototype operates as described. It should offer an intuitive understanding from a user's perspective and an experience transcending static musical forms. The UI should be clearly labelled or naturally intuitive, with the *Max* for Live patches optimised for presentation mode. The UI must provide substantial control over the parameters of your *Max* for Live and *Ableton Live* set.

Video and Instructions: Address UX functionality, the type of listening experience provided, user engagement, and the exploration appeal in your video. Detail your strategy for refining the prototype in future iterations. Your user instructions should be clear and add to the overall usability of your system.

Embrace this creative journey to enhance your technical skills and broaden your understanding of the transformative potential of interactive music.

Notes

1 For more details on scripting in TouchOSC, see https://hexler.net/touchosc/manual/script
2 https://www.ableton.com/en/packs/max-live-building-tools/
3 A bpatcher object in Max. Is a subpatch with a visible User Interface (UI).
4 https://docs.cycling74.com/max8/vignettes/live_object_model
5 https://www.kadenze.com/courses/machine-learning-for-musicians-and-artists/info

12   Adaptive Music for Video Games

Overview of Tutorial Series

In this chapter, we will look at composing adaptive audio in gaming. The chapter will introduce vertical orchestration, horizontal resequencing, audio middleware *FMOD*, audio integration and *Unity*. Given that the gaming industry has the most established approaches for composing transmutable music, whether you aim to work in game composition or not, you will find new compositional approaches that can be transferred to other Transmutable Music works.

The first part of the tutorial includes composing a vertical orchestration controlled by an intensity parameter for a space-themed video game. Followed by an introduction to *FMOD* and *Unity*. We will cover the basics of *FMOD* and *Unity* and develop a prototype so you can test out your composition in a game environment.

The second part of the Tutorial involves composing music for the next levels in the spaceship-themed game using horizontal resequencing. We will examine composing and designing transitions, audio integration and *Unity* prototyping.

Finally, I will challenge you to compose your own adaptive music and sound design for a demo game, *Vikings Village*.

As we begin this tutorial series, I want to acknowledge two of the influences that have shaped how a couple of these lessons were crafted. *The 'Game Music Composition: Make Music for Games from Scratch* course on Udemy, created and taught by Karleen Heong, is a great resource that inspired me with a couple of tutorials here. More details about this course can be found at https://www.udemy.com/course/gamemusiccourse/

Also, a shout-out to Daniel Sykora's FMOD video series on *Viking Village Sound FX for Unity 5*. It's a bit dated now, but it was a pivotal guide when diving into *FMOD*, helping me build on my knowledge from using *Wwise*.

Tutorial 12.1 – Vertical Orchestration/Remixing (Layering and Blending): Building Suspense and Intensity

This tutorial will cover techniques for composing intensity and suspense for video games. We use compositional techniques, including layering, blending,

DOI: 10.4324/9781003273554-16

and adding variation. We will write four layers of intensity, then mix and prepare loops for our game prototype. The layers fade in and out based on the gameplay. In tutorial 12.2, we will design and implement these loops into *FMOD* and *Unity*. This tutorial assumes an intermediate level of knowledge in *Ableton Live*. For help, go to the Companion Website video tutorials. The completed *Ableton Live* project is available for download from the Companion Website.

Composition Brief

You are tasked with composing four 8-bar loops that progressively build in intensity. These loops will serve as the background music for a game where the player is aboard a spaceship. The player's goal is to explore the ship to find health packs and tools necessary to start the spacecraft.

1 Control Room (Layer 1): Create an atmospheric, spacey, chilled soundscape to entice exploration. Include randomised playback elements to maintain variability during extended stays.
2 Machine Room (Layer 2): As the player enters the machine room, introduce an ominous pad sound to build tension and suspense with an atmospheric, dark, and noisy backdrop.
3 Alien Encounter (Layer 3): Introduce a percussive layer when the player discovers an alien hiding spot, increasing the intensity further.
4 Boss Fight (Layer 4): Culminate the tension with staccato strings during the boss fight, where the player battles the alien boss to obtain the final component needed to start the spaceship.

Compositional Techniques for Building Suspense

I recommend reading Winfred Phillips's excellent tutorial 'Composing video game music to build suspense.'[1] This contains some interesting research on music and psychological responses.

TECHNIQUES TO EXPLORE

- Ominous Ambience Technique: Use ambient sounds with mechanical or dissonant tones in the mid-frequency range to create suspense.
- Tempo Manipulation: Gradually increasing the tempo can also raise the players' heart rates, adding to the suspense.
- Crescendo: Increasing the music's volume can transform tension into revulsion, amplifying the suspense.
- String Ostinato: Short, repeated bursts of strings effectively build suspense.
- Theremin and saw sounds.

For further guidance on these techniques, refer to Winifred Phillips' tutorial on composing video game music to build suspense, which provides detailed examples and studies supporting these methods.

Exercise 1 – Layer 1 Control Room – Atmospheric, Spacey, and Chill

You can follow along with the tutorial or create your own piece loosely based on the instructions. The first layer we will create will be a calming layer that also aids in inspiring the player to explore the Spaceship.

1 Open a new Live set and save it as 'SpaceVO.'
2 Change the tempo to 98 (or a similar tempo). Choose a key or follow along, and we will use G minor.
3 Compose a very simple ambient sound layer. We will add another melodic instrument to this layer soon in step 7, so keep it simple. You can find your own instrument or start with a 'Sunrise Wave', 'Wavetable' Instrument. You can find it in the search area of the browser or follow this path: Instruments/Wavetable/Ambient&Evolving/Sunrise Wave.
4 Add the following MIDI notes into an 8-bar clip in the session view: D for 4 bars, followed by G for 4 bars. This loop starts on the 5th interval and then goes back to the tonic (G) – this will help to create confusion about where the tonic is. Creating confusion is the best way to add suspense to your compositions.
5 Let us mix as we go (Multiscale composition). Add a compressor to the MIDI track, bring down the threshold (−32db), and increase the makeup gain (14db) until you are happy with the sound.
6 To make this sound very spacy and atmospheric, add some effects. Choose a Convolution reverb, '06 Bigger Spaces,' 'Q2396Church,' and add an auto filter slowly sweeping using an LFO. See Figure 12.1.

Adding a Second Layer

7 Next, add another instrument for the melodic sound, the 'Echo Bay,' 'Wavetable' preset.
8 If you prefer to compose using a MIDI controller, arm the MIDI track, turn on the metronome and play song notes until you find something that sticks. Remember, it needs to be simple, as more layers will come. Or you can manually enter MIDI notes.
9 I suggest writing a reasonably long melody piece and then using 8-bar fragments; try creating three different melodies. Remember to keep it simple as we are layering the sound. See Figure 12.2.

Figure 12.1 Screenshot of *Ableton Live.*

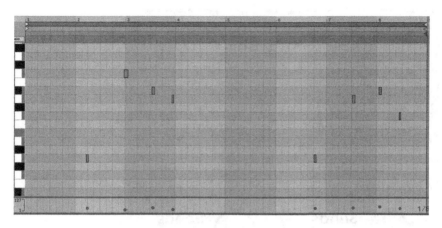

Figure 12.2 Screenshot of *Ableton Live.*

10 This choice of notes is in the key of G min. Set up follow-on actions for your melodic loops so you can test them out randomly. You may want to create some silence, too.
11 Let us blend the melody better into the background layer by adding an outer space delay of around 71% wet and a 'Pong Delay' device.
12 Tip - You could also play with the Oscillator position on the wavetable to transform the sound and control it with an LFO.
13 Create a new group for your two 'Layer 1' tracks. (Shortcut – Select both MIDI tracks by holding 'shift,' then press 'command' 'g' (Mac) or 'control' 'g' (PC).
14 Save your *Ableton Live* project if you haven't already.

Exercise 2 – Layer 2 Machine Room – Atmospheric, Noisy, and Dark

1 Add the 'Chill Outzone' 'sampler' instrument to a new MIDI track. You may want to choose another instrument that has a similar sound quality.
2 Start with improvised chords using the key you chose or G min if you are following along.
3 Record yourself and then edit to 8 bars. Try to keep it simple. Remember, a lot may be going on with sound effects and character themes. Or play or enter a G min chord (G2 root) for 8 bars using the 'chillout zone' instrument – this provides an excellent engine room sound for layer 2. You could also try a dissonant-sounding chord.
4 We will resample this MIDI clip into audio for our first audio loop. You should notice a slow attack if you listen to the start of this 8-bar loop. This means the loop will not loop seamlessly. To create a seamless loop, we will need to record 16 bars so we have room to edit the loop together.
5 Lengthen your MIDI chords to go for 16 bars instead of 8.

Figure 12.3 Screenshot of *Ableton Live*.

6 Move an empty audio track next to the 'Chill Outzone' MIDI track. Rename the audio track 'Layer 2 Audio'.

7 On the 'Chill Outzone' MIDI track, set the 'Audio to' instead of Master the 'Layer 2 Audio track.'

8 Next, on the 'Layer 2 Audio' track, change the "Audio from" to the 'Chill Outzone' MIDI track. See Figure 12.3.

9 Ensure you have a good gain level and it is not clipping. Arm the record button on the audio track. Then, press the round clip launcher to record around 18 bars of audio.

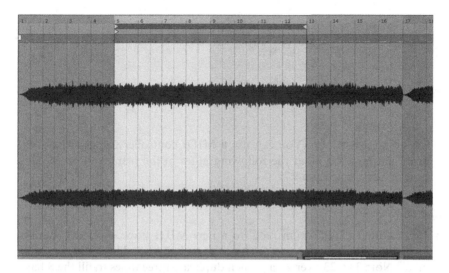

Figure 12.4 Screenshot of *Ableton Live*.

10 We can see from this screenshot (Figure 12.4) that the attack and release of the sound will not work to create a continuous, seamless loop. Find an 8-bar loop within the 16 bars that will work.
11 You can now crop this loop to 8 bars long. Loop the clip in clip view from bars 5 to 13.
12 Using a 'Utility' Device, I recommend adding 174% of stereo width to this sound.

Exercise 3 – Layer 3 Room with an Alien – Build Suspense

This will be our percussion layer to add tension.

Timpani

1 Add an empty 'drum rack' onto a new MIDI track and add the 'Timp Orch MPC 1.' You can search for this in the browser in samples.
2 A timpani drum is big and deep. However, let us make it deeper: Click on the 'Timp Orch MPC 1' sample in the 'drum rack,' and in the 'sampler,' move the transpose dial down an octave (−12 steps). See Figure 12.5.

Figure 12.5 Screenshot of *Ableton Live*.

Figure 12.6 Screenshot of *Ableton Live*.

3 Create an 8-bar loop. You can use a MIDI controller to play with some drum patterns. Or input the following notes in the piano roll on the first beat of bars 1, 3, 5, and 7.

Kick and Snare

4 Add another blank 'drum rack' to a new MIDI track. Add the 'Bunker kick' and Tronic Snare' samples into the rack. Next, add the 'Bunker kick' on 8th Note 332323 over 2 bars, then duplicate three times to fill the 8 bars. See Figure 12.6.
5 Finally, place the 'Tronic Snare' on the 3rd beat in each bar.
6 You can add variation to the velocity or timbre here, too. If all the audio is 8 bars, try adding some variation. In *FMOD*, we can add some pitch and filter variation later.
7 Create two more beat variations now after duplicating the first MIDI clip. Remove some beats, with the snare only on the 3rd beat of the 1st, 3rd, 5th, and 7th bars. Also, remove the 2nd and 4th beat in the 2-bar kick pattern. Finally, at the end of the 8 bars, try to create some variation by adding three 16th note snares.
8 The 3rd rhythmic variation will be snare and kick every beat 1, 2, 3, 4 of each bar, using 16th notes. You could also play around with the idea of skipping a beat.
9 Create a new group with the two drum tracks and add an 'EQ8' and a 'drum bus' device. Use the devices to get a sound you are happy with, or copy Figure 12.7.

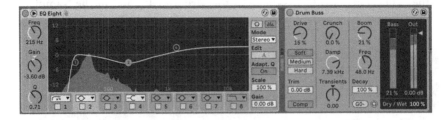

Figure 12.7 Screenshot of *Ableton Live*.

String Ostinato

An Ostinato can be understood as a small identifiable collection of notes playing in a repeating pattern. They can be simple, played with just one note repeatedly, or more complex. They can use the notes of a chord or scale, or any note you like. They are dramatic and rhythmic and are used in film scores, game composition, and orchestral pieces. An easy way to do a String Ostinato is to use the notes of your chords using 16th notes. We will do this in exercise 4.

Exercise 4 – Layer 4 Room, Alien Boss – Fearful

1 We will now create a string ostinato to add our final layer of suspense.
2 Start with a Cello instrument. The Staccato-style string will work the best for an ostinato. Select the 'Cello Solo staccato' instrument from the browser.
3 Use the same chords as our first layer from step 4 in exercise 1. First, use the D min chord for 4 bars and then change it to the G min chord for another 4 bars.
4 Use 16th notes, 'A, F, and D' in an arpeggiated pattern A, D, F, D. Then D, Bb, and G in the same pattern, as shown in Figure 12.7. Adjust the velocity of each of the notes for a bar, then duplicate each bar to make up the whole 8 bars. (Figure 12.8).
5 I will add a couple of extra chords to give the phrase more movement. I ended up using Dm Cm Gm Am.
6 Next, I changed a few of the high notes of each chord ostinato to give it a slight variation. (Figure 12.9).
7 Instead of programming the velocity to speed things up, you can use the random velocity with a range of = 38 in the velocity range in the clip view, then press randomise. You can keep randomising the notes until you like the performance and then tweak it.

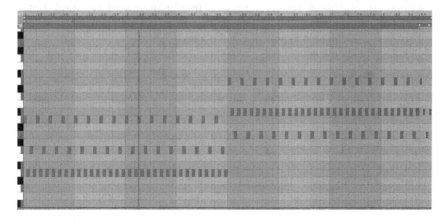

Figure 12.8 Screenshot of *Ableton Live.*

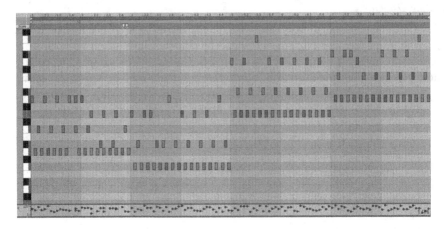

Figure 12.9 Screenshot of *Ableton Live*.

8 One way to create fear and suspense is to use intervals, minor 3rd, minor 2nd, and the tonic occurring rapidly. You might like to try this out.
9 To add more power to this layer, we will add three more instruments: 'violin,' 'brass,' and 'double bass section staccato' instruments.
10 Copy the 'cello' MIDI clip to a new MIDI track with a violin section staccato instrument. Transpose up two octaves and remove some or all of the MIDI notes in the 1st and 5th bars. Tweak the 'violin' MIDI clip to emphasise certain parts. Alternatively, you could compose new parts for the new string sections.
11 Next, add a 'double bass solo' instrument to a new MIDI track. Using your keyboard, find a simple pattern that adds a menacing sound. Or do the following: Starting at the 5th bar, enter the following 1 bar in length: G0, Bb0, Eb1, D1.
12 Finally, try the 'Swell Brass instrument' to add an unworldly sound. Enter a G note on the 1st and 5th bars, the first note shorter than the second.
13 Create a new 'layer 4 group' for the ostinato phrase.

Exercise 5 – Mixing

1 Tidy up your set with tracks, groups, and layers named and organised.
2 We have the following loops to make:

    a Layer 1 – 4 loops – 1 Sunrise wave loop and 3 Echo Bay variation loops.
    b Layer 2 – 1 loop
    c Layer 3 – 2 loops – 2 drum patterns
    d Layer 4 – 1 loop

3 We must mix everything based on how it will be played back in the game. Our layers will not be played separately. As they will only be built upon, we will mix for that format.

4 First, we will create a reverb send to glue all our sounds together in the same space.

5 Add the 'Convolution Reverb Pro' *M4L* device into your 'return A' track and delete the existing reverb. Select the Berlin power station 'type' in the 'Convolution reverb' to give it an industrial feel. We will leave the delay device in 'Return B'as is for this exercise as it is a quick demo.

6 Bring all the track's gain levels down to −12db if you haven't already. This way, we can create some headroom in our mix. Aim for the master to be peaking between −6db and −12db.

7 Add 'compression' and 'EQ Eight' to each individual track.

8 Think about each track's frequency range and remove unwanted or unneeded frequencies using the 'EQ eight.'

9 Experiment with the right compression, boom, and drive using the 'drum bus' on the layer 3 group, adjusting the volume as you go.

10 Compress each track. You can do the EQ and compression simultaneously for each track, too. It's up to you.

11 Mix a good level between the string section. Add reverb to the group via the send.

12 Add reverb via the sends to all the tracks.

13 Experiment with the delay send on the drums to add something more to the track.

14 Listen to the mix and try playing it back in order, adding layers. Start with layer 1, then add layers 2, 3, and 4. Tweak the gain of each layer and track until you are happy with it.

15 Add a Master chain on the master track and get a good level. Mastering can also be done within *FMOD*.

16 Export your 8 loops, as per step 2 of this exercise.

17 Save your project if you haven't already.

Tutorial 12.2 – Prototyping and Audio Integration: Intro to *FMOD* and *Unity*

Overview

In Tutorial 12.2, we take the music loops created in the first tutorial and learn how to implement them into *FMOD*, a sound effects engine for video games. This lesson will introduce you to the fundamental techniques of using *FMOD*, particularly in applying the concept of Vertical Orchestration. You'll learn to set up a parameter that controls the blending of different music layers based on the game's intensity.

Then, we'll move on to *Unity*, a popular game development platform. You'll learn how to incorporate *FMOD* into a *Unity* project and use it to prototype a basic concept. This allows you to see and hear how your music and sound effects from *FMOD* work within an actual game setting. Parts of this tutorial have been adapted from part of Karleen Heong's Udemy course *Game Music Composition: Make Music for Games from Scratch.*[2]

Resources and Tools

The entire tutorial series, including this one, can be found on the Companion Website. There, you'll also find guidance on downloading and setting up *FMOD* and *Unity*, both of which are free to use.

Tutorial 12.3 – Sound Effects and Variability in *Unity*: The *Viking Village* Project

What You'll Do

In Tutorial 12.3, we'll begin by setting up a game project in *Unity* called *Viking Village*. You'll be guided through crafting sound effects (FX) in *FMOD*, emphasising adding variation to make the game audio more dynamic and realistic. After creating these varied sound FX in *FMOD*, we'll move on to incorporating them into the *Viking Village* game, which is a premade *Unity* project. This tutorial has been adapted from Daniel Sykore's tutorial series[3] for FMOD.

Integration and Testing

Once the sound FX have been designed and are ready, you'll learn how to seamlessly integrate your *FMOD* project into *Unity's Viking Village* game. This integration will involve placing your new sound FX within the game's environment. The final step is to test the sound FX inside the game to ensure it works correctly and enhances the overall gaming experience.

Resources

For a comprehensive learning experience, this tutorial series, including the necessary *Unity* project and sound FX, is available on the Companion Website. Here, you'll find all the resources needed to follow along with the tutorial and successfully add immersive audio to your *Unity* game project.

Tutorial 12.4 – Horizontal Resequencing for Adaptive Audio: *FMOD*

Overview

In Tutorial 12.4, we take a deeper dive into the capabilities of *FMOD* with a focus on Horizontal Resequencing. This technique is about managing musical compositions in a way that allows for dynamic changes during gameplay. Building on the spaceship game concept from Tutorial 12.1, we now add four states the player can experience at any point in the game.

What You'll Learn

You'll learn how to structure your *FMOD* events to ensure smooth transitions between these states as they get triggered within the game. We'll cover more sophisticated uses of *FMOD* parameters to control these transitions. You'll also discover methods to enhance the variation in your music tracks and create looping audio that keeps the player engaged without noticeable repetition.

Advanced Techniques

This tutorial introduces you to advanced parameter control. It demonstrates simple yet effective ways to implement transitions in your music for a horizontal resequencing format, which is crucial for maintaining an immersive game environment.

Resources

This tutorial series is on the Companion Website, accompanied by supporting audio files and an *Ableton Live* Project file. These resources are designed to help you grasp and apply the concepts to your game audio and Transmutable Music Projects.

Tutorial 12.5 – Horizontal Resequencing and Audio Integration: *FMOD* and *Unity*

Final Steps in the Series

This last tutorial provides an overview of the nuances of creating fluid transitions within complex musical compositions and how to effectively implement these in *FMOD*. It moves beyond *FMOD* to show ways to integrate your completed audio project into a *Unity*-based Proof of Concept game.

Integration Techniques

You'll gain hands-on experience linking *Unity's* interactive elements to the *FMOD* project using the *FMOD* integration tools. This process involves setting up triggers in *Unity* that respond to game events with appropriate audio changes, enhancing the overall gaming experience.

No Coding Needed

This tutorial is designed to be accessible to those with little to no coding experience. It focuses on the principles of adaptive audio in gaming and how to apply transmutable music composition techniques, offering a gateway into game audio design.

Final Exercise Challenge

The culmination of this tutorial is a challenge in which you'll create an adaptive audio score for the *Viking Village Unity* Project. This exercise will test your understanding of the series and your ability to apply these techniques practically.

Where to Find Resources

All necessary materials, including project files and detailed tutorial instructions, are provided on the Companion Website, equipping you with everything needed to complete this comprehensive learning experience in adaptive audio for games.

Creative Exercise – Composing Adaptive Audio and FX for the *Viking Village*

Compose and implement your own adaptive music score and sound effects into the Vikings Village game. Compose the adaptive music exploring vertical orchestration and horizontal resequencing techniques. You should aim to get your transitions to work seamlessly.

Using a DAW, FMOD, and Unity, you can practice what you have learned from the tutorials in this chapter and integrate your composition into a demo game.

The details for this creative exercise, along with supporting resources, can be accessed via the Companion Website.

Notes

1 https://www.gamedeveloper.com/audio/composing-video-game-music-to-build-suspense-part-1-ominous-ambience
2 https://www.udemy.com/course/gamemusiccourse
3 https://www.youtube.com/watch?v=KkQ89ZXv5sQ

Section 5

Conclusion, Companion Website, and Glossary

13 Conclusion

Conclusion

This book explores the role of interactive technologies and their influence on music creation. It looks at music's ability to adapt to technological change and the opportunities and challenges of current interactive technologies on traditional music making practices.

The book provides the following tools and approaches in its discussion of how interactive technologies can be used to develop new forms of music:

- Defining 'Transmutable Music' as an umbrella term to represent music that can change and adapt to data, thereby offering new experiences.
- Discussing the relationship between Transmutable Music and traditional concepts such as form, structure, process, and time.
- Presenting the four components of a Transmutable Music system: experience, content, musical architecture, and control system.
- Proposing a model for the musical design of Transmutable Music.
- Offering practical guides and basic tutorials for creating Transmutable Music focusing on experience, variability and transmutability.
- Offering criteria for evaluating Transmutable Music in comparison to static forms, with the hope of assisting artists in innovation within this field.
- Providing two advanced tutorial series on Mobile Music Making and Adaptive Audio in Video Games to assist musicians and composers in practical approaches to making Transmutable Music.

Music and technology go hand in hand through life as we know it. However, we have reached a standstill: recorded music in its current form cannot meld with the new technologies of our time. This is reflected in the central concept of this book. That change is the only constant and everything is in a constant state of flux. Static recorded music is fixed in time. As technology starts to represent the state of constant flux in new and improved ways, our experiences of recorded music might need to catch up.

The implications of interactive technologies in music creation, production, distribution, and consumption are discussed. Despite the digitisation

DOI: 10.4324/9781003273554-18

of music, most music still needs to be released in static forms like singles or albums. This seems limiting, considering the potential for digitisation to transform recorded music into something that can change with each playback.

The book posits that if recorded music is now digitised and can be represented as data, it has the potential for complete transformation by interactive technologies. However, parts of the industry and artists are still rooted in static forms due to institutionalised processes and societal expectations. A shift in perception is required to move beyond, acknowledging that music changes every time it is performed and that digitisation could revive the dynamic nature of music experienced before analogue recording technology.

Copyright and royalty models desperately need an overhaul to reflect current and future consumer and artistic trends. The speed at which these changes are being developed strongly indicates the pace of music innovation within the mainstream. In the era of effortless duplication ushered in by AI and digital technology, Transmutable Music stands out as inherently uncopiable. While it is possible to replicate particular renditions or attempt to mimic its system, the dynamic nature of the music thwarts direct copying. This further challenges current copyright models with no current alternative to copyright a Transmutable Music system other than a patent.

Throughout the book, I emphasise that Transmutable Music will not replace traditional music forms and formats. These traditional, established fixed forms are the mainstream and will not change anytime soon. I see Transmutable Music as more of a branch of music that might begin to grow. Music, after all, is so much more than its recorded form. As well as an art form in its own right (which is arguably completely undervalued at this current time), music is at the heart of the creative industries. Video games, movies, TV series, documentaries, and advertisements depend on music to produce successful outputs. Music is engrained in almost every aspect of our lives. It helps us through the hardest and happiest of times. It becomes part of our memory and nostalgia. It has been used by physicists and philosophers alike to help discover and explore the universe we live in. Learning music helps our brains develop new synapses, assisting with intelligence. I could go on; however, my point is that music is much more than a single form or genre. Transmutable music is another pathway to explore at this time, with the potential for transformative innovation.

Traditional music roles and processes are also converging. DAWs blur the methods of composition, recording and mixing, converging into a multiscale compositional approach. Artists compose the sound they compose with, blurring the idea of an instrument. New remix applications, interactive, generative, contextual, AI, and autonomous music forms further challenge the traditional notion of an instrument. Remix culture has blurred the roles of consumers and producers. What will the future hold for mainstream music? Will Transmutable Music provide a pathway to revolutionise traditional roles and processes in the future of music?

The skills needed for Transmutable Music production differ from traditional music making, requiring knowledge in transmutability, data science, programming, interaction design, user experience, and techniques to produce variability. It calls for a collaborative effort among musicians, technologists, and researchers to forge new paths in the field of Transmutable Music, harnessing the power of modern interactive technologies to meet and exceed evolving consumer expectations.

The book concludes that more research is needed into the audience experience for Transmutable Music to gain popularity. It suggests that the success of Transmutable Music might depend on how well the audience understands the changes in the music. The potential for Transmutable Music to offer new narrative forms and storytelling in music is also noted, offering multiple perspectives and concepts in the same work.

We are in a time where binary concepts in society are blossoming into accepted diversities. Technological advancement is exponentially ever-expanding. Could our songwriters have found a way to tell their stories from new perspectives, thereby challenging the binary views of the past?

The future of Transmutable Music is promising, with the potential to change how audiences experience music, requiring new platforms and a cultural shift in understanding and appreciating its dynamic nature.

Transmutable Music is an innovative way for artists and musicians to transform artistic processes and align them with current knowledge. It has emerged as a powerful force in an era increasingly defined by the 'data rush.' This rush refers to the datafication of human experience, rapidly infiltrating our personal lives and society as a whole. Transmutable Music has the potential to spark conversations and bring attention to ethical considerations, privacy concerns, and the impacts of surveillance capitalism. As Marshall McLuhan noted in 1964, 'I believe that art, at its most significant, is a DEW line, a Distant Early Warning system that can always be relied on to tell the old culture what is starting to happen to it' (McLuhan, 1964). This viewpoint highlights the transformative potential of Transmutable Music in today's data-driven world.

Reference

McLuhan, M. 1964. Understanding Media. *DEW-Line Newsletter*.

14 Companion Website

Chapter 14 introduces the Companion Website for the book. It is an essential online resource for readers. This website provides access to a host of supplementary materials intended to bolster the book's content, including software links, FAQs, and detailed tutorials. Structured into 'resources' and 'tutorials and exercises,' it covers practical components from chapters 3 to 6 and 10 to 12, supporting readers with exercises in design, user experience, and advanced audio programming techniques. To access the full suite of materials, readers must register their book purchase via the provided URL. This website is a crucial tool for enhancing the interactive music-making experience.

When you purchase this book, you can access its online resources. This includes helpful content such as links, a Frequently Asked Questions section, software, tools, and tutorials. All these resources are designed to enhance your understanding of the book's content and to help you better grasp and apply the methods it presents.

To obtain a link to the site, visit url and follow the instructions for registering your product. The Companion Website contains the following support materials.

The support materials are separated into two sections: 'resources' and 'tutorials and exercises.'

Resources

Information on Software requirements: Links to download the required software for the tutorials, trial information, and special discounts or offers.

FAQ: This page will list common issues regarding utilising the resources and tutorials.

Chapter 4

1 'Is this Your World' Max Patch
2 *One Drop* Max Patch

DOI: 10.4324/9781003273554-19

Chapter 5

1 'Is this Your World' Max Patch
2 *One Drop* – Copy of prototype
3 *OneDrop* Ableton set
4 *Semantic Machine* launch information

Chapter 6

1 Accessing a data set in Max, adding patch names and numbers
2 Accessing data from the internet via a weather API
3 Markov chain for drums device in Max4Live

Tutorials and Exercises

The tutorials featured in this book and the website include instructions, resources an,d/or videos.

Chapter 3

1 Exercise 3.1 Crafting Your Blueprint for a Transmutable Music Work

Chapter 4

1 Exercise 4.1 – Musical Design and User Experience
2 Exercise 4.2 – Conduct your own User Testing.
3 Exercise 4.3 – User Experience Evaluation and Discussion

Chapter 5

1 Tutorial 5.1 – Generative Work Using Variation.
2 Tutorial 5.2 – Composing an 18-loop Work (Phrase Level)

Chapter 6

1 Tutorial 6.1 – Chance and MIDI: Utilising Randomness in Max4Live

Chapter 10

1 Tutorial 10.1 – Working with *TouchOSC*
2 Tutorial 10.2 – Simple Mapping to Musical Parameters (MIDI)
3 Tutorial 10. 3 – Designing a Template / GUI with *TouchOSC* (MIDI)

Chapter 11

1 Tutorial 11.1 – Accessing Accelerometer data in *Max4Live* (OSC) – working with LOM
2 Tutorial 11.2 – Developing Simple Models with Decision Trees in Max4Live (OSC)
3 Tutorial 11.3 – Intro to Machine Learning using Accelerometer Data in Max4Live
4 Creative Exercise – Crafting Your Interactive Music Prototype with Mobile Integration

Chapter 12

1 Tutorial 12.1 – Vertical Orchestration (Layering and Blending): Building Suspense and Intensity.
2 Tutorial 12.2 – Prototyping and Audio Integration: Intro to *FMOD* and *Unity*
3 Tutorial 12.3 – Sound Effects and Variability in *Unity*: the *Viking Village* Project
4 Tutorial 12.4 – Horizontal Resequencing for Adaptive Audio: *FMOD*
5 Tutorial 12.5 – Horizontal Resequencing and Audio Integration: *FMOD* and *Unity*
6 Creative Exercise – Composing Adaptive Audio and FX for the *Viking Village*

15 Glossary

Adaptive Music Music that changes in response to real-time events or user interactions, often found in video games.

AI (Artificial Intelligence) The simulation of human intelligence in machines programmed to think like humans and mimic their actions.

Album Apps Digital applications that offer interactive and visual elements alongside audio content, transforming traditional album experiences.

Analytical Systems Systems that analyse data to extract insights and patterns.

API (Application Programming Interface) A set of rules and protocols for building and interacting with software applications.

AR (Augmented Reality) An interactive experience where the real-world environment is enhanced by computer-generated perceptual information.

Audio Analysis The examination and processing of audio to extract meaningful information.

Audio Middleware Software that acts as an intermediary between audio creation tools and game development platforms to manage and implement game audio.

Bpatcher An object in Max/M4L that allows encapsulation of a patcher within another patcher, often used for creating complex user interfaces and modular designs.

CD-ROM A pre-recorded optical compact disc that contains data accessible to, but not writable by, a computer for data storage and music playback.

Chaos Theory A branch of mathematics focusing on the behaviour of dynamical systems that are highly sensitive to initial conditions.

Classification The process of predicting the category to which a data point belongs.

Cognitive Psychology The study of mental processes such as 'attention, language use, memory, perception, problem-solving, creativity, and thinking.'

Componium An automated instrument that can produce variations of musical compositions.

Controllers Devices or interfaces that allow users to interact with the music system, such as GUIs, game controllers, or sensors.

Copyright A legal framework that grants the creator of original work exclusive rights to its use and distribution.

Creative Commons A licensing system that enables creators to specify the permissions for using their works.

Data Classification The organisation of data into categories that are meaningful for the system.

Data Mapping The process of linking control messages to specific parameters within a software environment.

Data Transformation The process of changing the form or structure of data to fit the needs of the system.

Data-driven Age   An era where data collection and analysis play a crucial role in how we understand and interact with the world.

Decision Stump   A simple decision tree used in machine learning with only one decision for classifying data points.

Deterministic   Systems or processes where the outcome is determined by the initial conditions, without any random elements involved.

Digital Era   The current period characterised by the widespread use of digital technology.

Digitisation   The process of converting information into a digital format.

Electronic Composition   The creation of music using electronic instruments or computer technology.

Extended Reality (XR)   An umbrella term encompassing augmented reality (AR), virtual reality (VR), and mixed reality (MR).

FM Synthesis   A method of sound synthesis using frequency modulation.

FMOD   An audio middleware solution used to create and implement adaptive audio for video games.

Free-Form Play   A mode of play in music games that is open-ended and not structured by predetermined sequences or scores.

Generative Music   Music that is algorithmically generated, often through a mobile app, allowing for a unique listening experience each time.

Gildo Method   A method developed in the early 11th century by Guido of Arezzo to teach chant and singing.

Glissandi (singular: Glissando)   A glide from one pitch to another, commonly used in music.

Google Maps API   A set of application programming interfaces provided by Google for incorporating maps and geographic data into websites or apps.

Graphical User Interface (GUI)   A visual interface that allows users to interact with electronic devices through graphical icons and audio indicators.

Gyroscope   A device that measures or maintains orientation, typically used in mobile devices.

Heraclitus (535 – 473 BC)   An ancient Greek philosopher known for his doctrine that change is central to the universe.

Horizontal Resequencing   A method of adaptive music composition where different segments of music are arranged in a sequence that can change dynamically.

Human-Computer Interaction (HCI)   The study and practice of designing, developing, and using computer technology, focusing on the interfaces between people and computers.

ICRAM   Institute for Research and Coordination in Acoustics/Music, a French institute dedicated to the promotion of contemporary music and sound art.

Interaction Design   The design of interactive products and services focusing on user's experience and interaction.

Information Theory   A branch of applied mathematics and electrical engineering involving the quantification of information.

Interactive Art   A form of art that involves the spectator in a way that allows the art to achieve its purpose.

Interactive Composition   A composition method where live interactions with a computer music system shape the music in real-time.

Interactive Music   Music that allows real-time interaction or manipulation by the listener, often through digital interfaces or interactive technology.

Interactive Technologies   Technologies that allow for two-way communication or interaction between the user and the system.

Johann Philipp Kirnberger   An 18th-century German musician and composer who developed methods of musical composition.

Live Object Model (LOM)   A framework within Max for Live that provides structured access to components of Ableton Live.

M4L (Max for Live)   A visual programming environment that runs within Ableton Live.

Machine Learning   A field of computer science using algorithms to enable computers to learn from and make decisions based on data.

Max4Live   A version of the Max visual programming language that integrates with Ableton Live.

MIDEM   An annual trade fair for the music industry.

MIDI (Musical Instrument Digital Interface)   A technical standard for connecting electronic musical instruments, computers, and other audio devices.

Mobile Music   A field focusing on musical interaction in mobile contexts, using portable technology.

Model   A representation, often mathematical, of a system or concept.

Multitrack Recording   A method of sound recording allowing for the separate recording of multiple sound sources.

Music Consumption   The process of experiencing music through various means such as listening, attending performances, or streaming.

Music Experience   The act of engaging with music, which can be shaped by various factors including the format, technology, and interactive elements involved.

Music Industry   The sector of the economy that creates, performs, promotes, records, and sells music.

Music Information Retrieval (MIR)   A field of research that develops methods and systems for retrieving information from music.

Music Interaction   A field of study focusing on the interaction between users and musical systems or instruments, often involving technology.

Music Production   The process of creating music, which can include composition, recording, mixing, and mastering.

Music System   A structured arrangement of musical elements and technologies that interact to produce music.

Music Video Game   A genre of video games in which the gameplay involves interaction with a musical score or individual songs.

Musique Concrète   An experimental form of music composition that uses recorded sounds and tape manipulations to create a musical collage.

Non-deterministic   Systems or processes that include randomness, making it impossible to determine the outcome with certainty from the initial conditions.

Nonlinear Dynamics   The study of systems that are not linear, meaning the output is not directly proportional to the input, often leading to complex behaviour.

Objective Choice   Decisions made based on algorithmic or rule-based selection, rather than human preference.

Ontology   A branch of metaphysics dealing with the nature of being, or a set of concepts and categories in a subject area or domain that shows their properties and the relations between them.

OSC (Open Sound Control)   A protocol for networking sound synthesisers, computers, and other multimedia devices for purposes such as musical performance or show control.

Participatory Culture   A culture where the audience actively participates in the creation and remixing of content, as opposed to passively consuming it.

Post-Digital   A cultural period where the distinction between digital and non-digital is increasingly blurred, influencing all aspects of society including the arts.

Printed Music   Musical compositions that are written and printed on paper.

Protokol   Software used to monitor and log OSC and MIDI data, useful for debugging and configuring connections between devices.

Pure Data (Pd)   An open-source visual programming language for creating interactive computer music and multimedia works.

Qualitative   Data that is descriptive and characterises rather than measures.

Quantisation   In music video games, it refers to the alignment of music elements to a grid to simplify player interaction.

Quantitative   Data that can be expressed as numerical values and can be measured and quantified.

Reactive Music Apps   Applications that produce music in response to user behaviour or environmental factors in real-time.

Regression   A set of statistical processes for estimating the relationships among variables.

Remix Competitions   Contests where individuals are invited to create remixes of songs, often with the possibility of winning prizes and recognition.

Remixing   The process of taking existing musical elements and recombining them to create a new piece of music.

RjDjRNBO   A mobile app that personalises music by using the built-in sensors of mobile devices to adapt the music to the listener's environment.

RNBO   A patching environment that allows for the creation and export of Max patches for use in various applications, including web and hardware devices.

Royalties   Payments made to rights holders (like songwriters and musicians) for the use of their music by others.

Sample Rate   The number of samples of audio carried per second, measured in Hertz.

Semantic Audio   Techniques used to extract or assign meaning to audio, such as classifying a sound as 'sad' or 'upbeat.'

Semantic Web (Web 3.0)   An extension of the World Wide Web through standards set by the World Wide Web Consortium (W3C) that enable users to create, share, and connect content through search and analysis.

Sonify   To convert data into non-speech audio signals.

Spatial Audio Technologies   Technologies that enable the creation and playback of immersive audio experiences beyond traditional stereo, including 3D and binaural sound.

Static Formats   Traditional methods of music distribution and consumption that offer a fixed, unchanging version of musical performances.

Stems   Individual components of a music track (e.g., vocals, drums, bass) used for remixing.

Streaming Services   Online platforms that deliver media content like music and video to consumers on-demand over the internet.

Subjective Choice   Decisions made based on personal preference, opinion, or individual interpretation.

Synchronicity   The simultaneous occurrence of events that appear significantly related but have no discernible causal connection.

System   A set of interacting or interrelated elements that act according to a set of rules to form a unified whole.

TouchOSC   An application for iOS and Android devices that turns the device into a custom control surface for sending MIDI or OSC commands wirelessly.

Transmutable Music   Music that evolves and changes, influenced by data, moving beyond traditional static playback methods.

Transmutable Music System   An ecosystem within which music adapts and changes according to input data or interactions.

Unity   A cross-platform game engine used for developing video games for web plugins, desktop platforms, consoles, and mobile devices.

User Experience Design   The process of creating products that provide meaningful and relevant experiences to users.

User Interaction　The ways in which the user engages with the Transmutable Music system, which can be direct or indirect.

User Testing　The process of testing the functionality and user interface of a product or service by observing real users as they attempt to complete tasks.

Vertical Orchestration　A method of adaptive music composition where layers of music can be added or removed to change the texture or intensity without altering the harmonic structure.

VR (Virtual Reality)　A simulated experience that can be similar to or completely different from the real world.

Weather API　Application Programming Interface that allows retrieval of weather data and forecasts, often used in apps requiring environmental context.

Web 2.0　The second stage of development of the internet, characterised by the change from static web pages to dynamic or user-generated content and the growth of social media.

Web 3.0　The evolution of web utilisation and interaction, which includes altering the web into a database.

Web Audio API　A high-level JavaScript API that allows developers to create and manipulate audio content in web browsers.

WebPd　An open-source compiler for the Pure Data audio programming language, allowing for the execution of .pd patches on web pages.

Wekinator　An open-source software project that allows users to build new interactive systems by mapping inputs to outputs using machine learning.

Wwise　An audio middleware software designed to integrate complex audio functionalities into video games.

XR (Extended Reality)　An umbrella term encompassing augmented reality (AR), virtual reality (VR), mixed reality (MR), and other immersive technologies that merge the physical and virtual worlds.

Index

Note: Page numbers in *italic* refer to figures, page numbers in **bold** refer to tables

Printed in the United States
by Baker & Taylor Publisher Services